Behavioral Science

MOSBY'S USMLE *step 1* REVIEWS

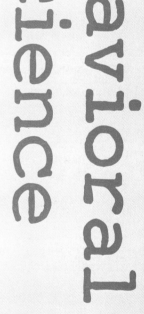

Steven Cody, Ph.D.

Associate Professor,
Department of Psychiatry,
Marshall University School of Medicine,
Huntington, West Virginia

St. Louis Baltimore Boston Carlsbad Chicago Naples New York Philadelphia Portland
London Madrid Mexico City Singapore Sydney Tokyo Toronto Wiesbaden

A Times Mirror
Company

Vice President and Publisher *Anne S. Patterson*
Editor *Emma D. Underdown*
Developmental Editor *Christy Wells*
Project Manager *Dana Peick*
Production Editor *Jeffrey Patterson*
Manufacturing Supervisor *Karen E. Boehme*
Book Designer *Amy Buxton*
Cover Designer *Stacy Lanier/AKA Design*

Copyright 1996 by Mosby–Year Book, Inc.

All rights reserved. No part of this publication may be reproduced, stored in a retrieval system, or transmitted, in any form or by any means, electronic, mechanical, photocopying, recording, or otherwise, without written permission of the publisher.

Permission to photocopy or reproduce solely for internal or personal use is permitted for libraries or other users registered with the Copyright Clearance Center, provided that the base fee of $4.00 per chapter plus $.10 per page is paid directly to the Copyright Clearance Center, 27 Congress Street, Salem, MA 01970. This consent does not extend to other kinds of copying, such as copying for general distribution, for advertising or promotional purposes, for creating new collected works, or for resale.

Printed in the United States of America
Composition by Graphic World, Inc.
Printing/binding by Plus Communications

Mosby–Year Book, Inc.
11830 Westline Industrial Drive
St. Louis, Missouri 63146

Library of Congress Cataloging-in-Publication Data

Cody, Steven.
   Mosby's USMLE step 1 reviews—behavioral science / Steven Cody. —1st ed.
      p.   cm. — (Ace the boards)
   Includes bibliographical references and index.
   Added title page title: Ace behavioral science.
   1. Medicine and psychology—Examinations, questions, etc.
2. Human behavior—Examinations, questions, etc. I. Title.
II. Title: Ace behavioral science   III. Series.
   [DNLM: 1. Behavioral Medicine—examination questions.   2. Human Development—examination questions.   3. Mental Disorders—examination questions.   4. Ethics, Medical—examination questions.   5. Delivery of Health Care—economics—examination questions.   WB 18.2 C671m 1996]
R726.5.C62 1996
616.89′0076—dc20
DNLM/DLC
for Library of Congress   96-28844
                                                                                     CIP

ISBN 0-8151-1844-9 (IBM)
ISBN 0-8151-1487-7 (MAC)
96 97 98 99 00 / 9 8 7 6 5 4 3 2 1

PREFACE

Medical students preparing for the USMLE exams certainly have a challenging task. Over the course of their medical education, they have encountered a vast body of concepts and facts, any and all of which could show up in the questions they are asked. Obviously, some facts and concepts are more important or fundamental than others. The first task for students, then, is one of organizing the effort and identifying the core of what they need to know.

ACE THE BOARDS: Behavioral Science was designed to provide organization and focus for the student reviewing fundamentals of both basic and clinical behavioral science, with the emphasis on concise, pertinent information. The bulleted, colorized format, and numerous boxes and tables, provide organization and additional highlighting of essentials. Icons identify materials sharing a broader theme or focus. The page format facilitates adding your own notes and references.

Each chapter includes several review questions, providing immediate self-assessment. The answer sheets are conveniently perforated, so they can be removed and placed next to the questions when you are reviewing the answers. Explanations are provided for all the possible answers. This way, you know why the correct answer was correct, *and* why the other answers are incorrect. The computer diskette will give you additional opportunities to assess yourself with questions in USMLE format. The categories and tracking feature on the diskette allow you to assess exactly which areas to study.

The subject matter is organized into five broad areas. The *Foundations* section reviews the biology of behavior, psychosocial models, and the topics of assessment and measurement. The *Development* section is a lifespan review of age-related changes and issues. In *Behavioral Health,* the focus is on behavioral issues in caring for patients (e.g., the doctor-patient relationship) and on areas of clinical interest where behavioral factors are important (e.g., sleep, stress, and substance use). The *Psychopathology* section is a review of psychiatric disorders and their essential features. Finally, *Social Context* brings together the topics of health care delivery and economics, legal and ethical issues, and the broader perspectives of the social sciences.

Much, if not all, of this book has been shaped by eight years of feedback from successive classes of medical students at Marshall University, who have consistently challenged me to be concise, organized, and responsive to their needs for effective use of their time in the face of multiple demands. Their responses have guided a continuing process of revision and refinement of the written material provided in the classroom as outlines and learning aids. Those materials have been the foundation of this book.

In addition to using this and the other books in the *ACE THE BOARDS* series, it is important to apply generally important behavioral principles to your preparation for the challenge of the USMLE exams. Under stressful circumstances, it is *more important*, not less, to take care of yourself. Start early, and include in your schedule the sleep, nutrition, and exercise you need to get the most from yourself. Take occasional breaks during study sessions, since this will enhance recall. A study partner or group may help to keep you focused. Remind yourself periodically that your task is not to know everything; your goal should be to solidify the preparation your education has given you to think like a well-informed physician should think.

ACKNOWLEDGMENTS

I thank Emma Underdown at Mosby for the opportunity to take on this engaging task. I am especially appreciative of the efforts of Christy Wells to keep me on target and productive, and the patience of Natalie Gehl in shepherding me through the practical tasks.

I am greateful to all the medical students I've been privileged to teach over the past decade. A teacher could hardly ask for brighter or more motivated students, and could hardly help but grow in response to the challenge of working with them. The following medical students thoughtfully reviewed this manuscript:

Vishal Banthia
Maria L. Camarda
Marcus Eubanks
Felipe Gutierrez
John W. Hariadi

I owe a special debt to my friend and predecessor at Marshall, Dr. Danny Wedding, who has been my mentor as a medical educator and clinician. The current chairman of the Department of Psychiatry, Dr. Dan Cowell, continues a tradition of support for my role as director of psychiatry courses in the preclinical years.

Finally, I am pleased to acknowledge the most profound debt of all, to my wife Lyn and my daughter Stephanie, for their unfailing tolerance and encouragement.

Steven Cody, Ph.D.

Test-Taking Strategies

Suzanne F. Kiewit, M.Ed.

To perform well on the USMLEs, it is imperative that you begin with a **plan**. Preparation time is at a premium, so you will want to be as efficient and effective as possible by planning well.

MONTHS AHEAD OF THE EXAM
- Sit down with a blank calendar and block in your commitments: classes, final exams, scheduled events.
- Include time for activities of daily life: eating, sleeping, exercising, socializing, banking, maintaining your home, and so forth.
- The remaining time is available for study/review.
- Determine an orderly approach to the material you need to cover that fits your particular set of needs (e.g., subject-by-subject approach, systems approach, pathologic state approach).
- Assign the remaining time to content areas. This is done in various ways: material covered freshman year first, easiest first, least comfortable material first, detailed subjects last, whatever. Your plan should reflect your goal: to maximize your score.
- Establish a warm-up, which may consist of breaking the tension in major muscle groups (neck rolls, shoulder rolls, etc.), a quick visualization of you performing successfully, or a brief meditation. Practicing this warm-up routine before each of your study sessions will make it a familiar activity that helps you learn effectively, as well as take exams effectively.
- Designate time at the end of your study period for panoramic review. Depending on your needs, that might be a week or just several days before the exam.
- Plan for feedback on your efforts. Schedule time for answering questions on the material you are reviewing and for taking at least one mock comprehensive exam.
- Do the comprehensive exam midway through your study period so that you can refine your efforts to reflect the degree of your performance.

DAYS AHEAD OF THE EXAM
- Divide each day into thirds: morning block, afternoon block, and evening block.
- Consider the time of day that is most productive for you and do the most difficult or least favorite material at that time.
- Assign more blocks of study to those areas requiring the most review to reach a comfortable knowledge level.
- A popular way to use blocks is to pair subjects or materials. For instance, pair strong content with weaker content so that you are not always in the position of not knowing material (which would invite negative feelings or ineffectiveness). Or pair a conceptual subject with a detail subject, such as physiology with anatomy, so that you are not always doing the same kind of thinking (this invites positive effort).
- Use your most productive blocks of time for actual study/review. Use the nonpeak times for reinforcement of material covered or feedback by answering questions on material that you have covered.

Planning the Blocks
- Once you determine time allocations for each content area and an orderly approach that fits your needs and goals, you want to specify what you plan to do during each block.
- Be specific as to content area, material to study, and task; for example, MICRO: review chart on viruses; PHYS: answer questions on renal; and so forth.
- Each study block will last approximately 3 to 4 hours. To be most efficient and effective, plan to take a 5- or 10-minute break every hour. If you are having difficulty getting into the study mode, plan to study for 25 minutes, then take a 5-minute break. Reserve longer breaks for switches between subjects. Get up and move around on breaks.

BEFORE THE EXAM
Knowing about the USMLEs helps demystify them. In general, the USMLE exams are a 2-day, four-book examination. Approximately 3 hours are allotted per exam book. Each day you will complete one book of about 200 items in the morning and another book in the afternoon.

In each exam book, questions are organized by question type, not content. Specific directions precede each set of questions. Only two question formats are used: one-best-answer multiple choice items (which typically come first on the exam) and matching items (toward the end of the exam). Students have reported that one-best-answer items make up the bulk of the exam (70% to 75%) and negatively stemmed items make up only 10% to 15% of the questions (Bushan, Le, and Amin, 1995). Matching sets, which make up about 15% to 20% of the items, may include short leading lists or long leading lists of up to 26 items from which to choose.

From year to year there may be variations in the organization and presentation of both content and item formats. It would be wise to read the National Board of Medical Examiners' *General Instructions* booklet, which you will receive when you register. This booklet contains descriptions of content, item format, and even a set of practice items. Be certain you read this booklet and familiarize yourself with the questions.

You can further maximize exam performance by taking control. Adults tend to perform better when they feel that they have a measure of control. For the USMLE it is easy to feel out of control. You are told what time to arrive, where to go, what writing instrument to use, when to break the seal, and so on. You want to assume control of as many aspects as possible to maximize your performance:

STUDY Follow the sage advice of planning your work and working your plan to maintain satisfactory preparation with regard to study.

SLEEP Get a good night's rest. Sleep needs vary, but 6 hours is usually minimum. Try to get the appropriate amount of sleep that you require.

NUTRITION Maintain proper nutrition during both study time and exam time. Eat breakfast. Choose foods that help keep you on an even energy level. Eat light lunches on exam days. If you have a favorite food and can take it with you, treat yourself.

MEDICATION It may be cold, flu, or allergy season. Take no medications that may make you drowsy.

CLOTHING Heating and cooling systems are rarely balanced enough to suit everyone. Wear articles of clothing that can be added or removed as necessary. Strive for personal comfort.

READINESS Develop physical, mental, emotional, and psychologic readiness for the exam. Keep your thoughts about the exam and your preparation efforts running positively. You must believe that *you can do this!*

ARRIVAL Plan to arrive as close to the designated time as possible and still allow yourself sufficient time to check in. Keep to yourself so that other people's anxieties will not affect you. Take care of personal needs. Find your seat.

ACCLIMATION Settle in and get comfortable. Take several deep breaths...RELAX. A relaxed mind thinks better than a tense one—it's that old "fight-or-flight" syndrome. Do your warm-up routine to help you relax.

ATTENTION Pay attention to the proctor. Complete all identification material as required. Read all instructions carefully. Ask for clarification as needed. Do not open your booklet until told to do so. After you are told to break the seal, quickly glance through the whole test to see how it is set up and how questions are organized. Again, you want to take control of the situation. A quick review eliminates surprises and allows you to develop a plan.

DURING THE EXAM

Plan Your Approach

There are numerous approaches to answering questions. Answer questions in the order that appeals to you. Doing the easier ones first may give a psychologic boost; however, the ones you skip may stay on your mind and cloud your thinking.

Another approach is to answer each question in sequence. Start with the first one in the section with which you begin and fill in an answer for each question. Do not leave any blanks! The theory behind this is that if you spend any time at all on an item, you should mark your best response at that time and go on. If you are not certain of your choice, mark "R" in the test booklet for review and return to it later as you have time.

Some students plan to do the matching items first. Matching items are the last set of questions in the booklets. If you prefer matching items, this is a reasonable plan because it helps you get started with items about which you feel confident. It is also reasonable because matching items are not good items on which to guess if you run short of time. **You** must decide the order in which you want to do the questions.

Complete the Bubble Sheet or Answer Card Carefully

There are two schools of thought on this matter. One is to fill in the bubble sheet **item by item** as you go. This method minimizes transcription errors. The other method is **block transfer.** Complete a logical chunk of questions in the test booklet (one or two pages) and then transfer responses to the bubble sheet. Be sure that the last question number on the page is the last numeral you blacken. This method saves time and offers a mini–mental break at the end of each block. Such minibreaks help decrease fatigue during a long exam. Choose the method that will work for you and *practice* it as you take prep questions.

Budget Your Time

If only the allotted time and the number of questions were considered, you would have approximately 54 seconds per question. Obviously, some questions may go more quickly and balance out the ones that take longer. To keep track, you need a pacing strategy. A good strategy is to establish checkpoints at 30-minute intervals. When you overview the booklet, circle the numerals corresponding to where you should be at 30 minutes, 60, 90, 120, and so on. For example, if you have 200 questions on a 3-hour exam, you should be at question number 33 at 30 minutes. As you complete the exam, check your time at the circled items. This technique keeps you from watching the clock too much, yet permits multiple opportunities to adjust your pacing.

If you find yourself spending too much time on any one question, select your best choice at that time, mark an answer, and "R" it for later review. The point is to keep going. Laboring too long on one question limits you from responding to other items you may know well. Remember, controlling your time helps you maximize your points.

ANSWERING THE QUESTIONS

- **Read and *understand* the stems and alternatives.**
  The most frequent error made on exams is misreading or misinterpreting the various aspects of a question. The **stem** is the introductory question or statement. The **alternatives** are the options from which you select the one best response. To encourage reading and understanding, use a process.
- **Follow a process to answer questions.**
  1. Quickly read the stem.
  2. Quickly read the options. (Combined, the first two steps create a preview of the item.)
  3. Carefully read, underline, and mark the stem in a timely fashion.
     - Selectively underline key words and phrases.
     - Pay attention to nouns, verbs, and modifiers.
     - Circle age and gender.
     - Note data in telescopic form (e.g., ↑ BP).
     - Graphically represent material if it helps you to understand (e.g., diagram the renal tubule to answer a question about reabsorption).
  4. Carefully read each alternative. Mark as appropriate.
- **Consider each alternative as one in a series of true, false, or not sure (?) statements.**
  Read each alternative. Rather than slashing out the ones you eliminate, work with each one and designate it as **true, false,** or varying degrees of **true/false/?.** This marking strategy requires you to make judicious decisions about alternatives relative to the stem. It also provides a record of your original thinking, which will save you re-thinking time if you need to reconsider a question. Practice this strategy on preparation questions so it becomes second nature.
- **Avoid premature closure.**
  Sometimes you may read a question and anticipate a response. Such a reaction helps focus your attention. However, be sure to read *all* the options so that you are selecting the *best* response. In one-best-answer multiple choice questions, there is one *best* and several *likely* responses. Avoid being misled; consider all the alternatives.
- **Be leery of negative stems.**
  Negative stems require shifting to a negative thinking mode to determine which alternatives are not correct. You can avoid this shift by using this strategy:
  - Circle such words as *except, least, false, incorrect, not true* to raise your awareness of them.
  - Cross out the negative and read the stem as though it were a positive.
  - Mark each option as T/F/?. The F option will then be the appropriate choice.
- **Keep your original answers.**
  To change or not to change answers is a difficult decision. The answer depends on a person's previous history. If you are the kind of student who, if you change answers, changes them from wrong to right, then selectively changing answers may be worthwhile. If, on the other hand, your past experience has been to change right answers to wrong answers, selectively changing answers is probably not a good idea. Good performers change answers, but only if they have reason, such as acquired insight or discovery of misreading or misinterpretation.

- **Maintain an even emotional keel.**
  If a question upsets you, calm yourself. Take several deep breaths. Tell yourself, "I can do this!" Give yourself a mental or physical break. Pay special attention to the next two or three questions after a bout of emotional uneasiness. It is possible to miss items when attention is diffused.

THINKING THROUGH QUESTIONS

- **Use logical reasoning and sound thinking.**
  - Read the item carefully. After careful reading, ask "What is this question really asking?" Restate it so that you know what is being asked.
  - Engage in a mental dialogue with the question. Talk to yourself about what you do know. Always start with what **you** know. Verbalize your thinking.
  - If a diagram or graphic representation is included, orient yourself to it **first** so that the options do not lead your thinking.
- **Use information found within the questions themselves to help you answer others.**
  There will not be "gimmes" on a nationally standardized exam. However, there may be items or graphics that trigger remembrances.
- **Create a diagram, chart, map, or graphic representation of given information.**
  Material that is visually presented usually helps clarify thinking. Use selective, quick sketching as warranted.
- **Reason through information like a detective.**
  - Sift through the details (preview).
  - Determine the relevant information (selectively mark).
  - Put the clues together as in solving a puzzle (reason).
- **Read carefully and note key descriptors.**
  - Note words such as *chronic, acute, greater than, less than, adult, child*.
  - Attend to prefixes such as hyper-, hypo-, non-, un-, pre-, post-.
- **Analyze base words and affixes.**
  Studying a question at the word level may help you remember salient information. Look for base words or related words. Determine Latin or Greek word parts and use their meanings to assist you.
- **Consider similar options equally.**
  If you mark one alternative as "false" for a particular reason and another option is qualified for the same reason, it's probably "false" as well.
- **Trust the questions.**
  The questions are designed to determine if you have a working knowledge of the material. They are not written to trick you. You need to believe that your medical school curriculum and your study efforts prepared you for most of the questions.
- **Meet the challenge of clinical vignettes.**
  Longer, vignette items challenge you to discern the relevant from the irrelevant material. In doing so, you are given multiple clues to consider. To effectively handle the vignette item, follow this strategy:
  - Scan the stem and read the first several lines.
  - Skip to the end of the stem and read the last several lines.
  - Check the alternatives to narrow your focus.
  - Now that you know what the question is about, go back to the stem; read and mark what's important to your informed decision making.
  - Make good T/F/? decisions.
- **Reread your underlines and markings when you are down to two choices, at 50/50.**
  By the time you work through a stem and numerous alternatives, it is easy to lose the gist of the question. Checking your focus by rereading only the underlines ensures that you are answering the question being posed.

ANSWERING MATCHING ITEM SETS

Matching items are used to measure your ability to distinguish among closely related items. They require knowledge of specific sets of information. As you study, be alert to potential material that could be tested in this way.

Matching items can be formatted in two ways. **Short leading list matching** items include a set of five, lettered options followed by a lead-in statement and then several numbered stems. **Long leading list matching** items include a set of up to 26 lettered options, followed by a lead-in statement and then several numbered stems.

To efficiently deal with a short leading list item, consider it as an upside-down multiple choice item with the same repeated options. To handle it effectively, do the following:

- Scan the list; determine the topic.
- Read the lead-in statement; determine the focus.
- Quickly read the stem; then read and mark key words.
- In the left margin, create a grid with A, B, C, D, E at the top.

Test-Taking Strategies

- Make good T/F/? decisions about each stem, marking them in the grid. In this way you can see the pattern of your responses. Similarly, a grid with the item numbers can be drawn beside the leading list and responses marked there.

Handling long leading list matching items effectively requires some modifications in the process. It is not efficient to make T/F/? decisions about each option, so follow this strategy:

- Scan the list; determine the topic.
- Read the lead-in; determine the focus.
- Read a stem and generate your own response.
- Narrow the focus. Put a check mark by those related options in the long list.
- Read and mark specifics in the stem to differentiate among those alternatives you marked.
- Make good T/F/? decisions.

For each stem, mark the narrowed-list options with a different symbol (star, dash, etc.). Items are listed in logical order, alphabetically or numerically. When looking for an option such as "xanthinuria," do not start at the beginning of the list. Looking in the appropriate place saves valuable seconds.

TEST WISENESS

How a question is worded can often influence your response to it. Most clues about "test psychology" are a function of the way in which a question is worded—test constructors cannot rename body parts, drugs, diseases, and so forth. Being aware of the psychology behind the wording can often help you answer the test question.

Using techniques of test psychology to arrive at a correct answer has limited value on standardized exams because those who construct the exams are well aware of the use of these techniques. Nonetheless, being wise to these techniques of test psychology may add another point or two to your score, and they can also enhance your sense of control. Knowing these techniques provides additional strategies to employ should the question temporarily stump you.

The best way to take any exam is to be totally prepared with a strong knowledge base and personal test confidence. The following techniques should be used only if you have exhausted your knowledge base, eliminated all distractors, and cannot come up with the answer even with logical thinking and sound reasoning. Such techniques are **not** a substitute for knowledge, nor are they foolproof.

- **Identify common ideas or themes within the options and between the stem and options.**
- Circle repeated words in the options.
- Select the option with the most repeated words or phrases.
- Circle words repeated in both stem and options.
- Select the option that contains key words or related words from the stem. This is a stem/option repetition.
- **Beware of words that narrow the focus or are too extreme because they tend to be incorrect.**
  Circle such words as *all, always, every, exclusively, never, no, not, none.*
- **Options that are look-alikes are good candidates for exclusion.**
- **Note qualifiers that broaden the focus because they may be correct.**
  Circle words such as *generally, probably, most, often, some, usually.*
- **Identify antonyms or two opposing statements as potentially correct options.**
  Test constructors may use pairs of opposites, so this tip may lose its effectiveness.
- **Select the most familiar-looking option.**
  Always go from what you know. Alternatives with unknown terms may be likely distractors.
- **Select the longest, most inclusive answer.**
  This would include "All of the above" as a strong potential response.
- **In numerical items, knock out the high and low alternatives and select one in the middle that seems most plausible.**
- **In negatively stemmed questions, categorize responses; the one that falls out of the category is a likely candidate.**
- **Mark the same alternative consistently throughout the test if you have no best guess and cannot eliminate distractors.**
  Before the test, decide which letter (A, B, C, D, E) will be your choice. In this way, if you have given a question your best effort and cannot decide, mark your favorite response and move to questions that cover more comfortable material.

AFTER THE EXAM

- **Between sessions and overnight:**
  - Take a well-deserved break. Eat nutritionally.
  - If you feel the urge to study, study material that is comfortable, from a source with which you are familiar (e.g., personally developed study cards or your annotated review book).
  - If you discovered a recurring "theme," you might desire to consult that set of information.

- Do something pleasurable. Relax. Get a good night's rest.
- **After the final booklet:**
  - Recognize that this exam is a measure of what you know on a given day for a given set of information at a given point in time. Keep a reasonable perspective.
  - **Celebrate!**

References

Bushan V, Le T, Amin C: First aid for the USMLE Step 1, ed 5, Norwalk, Conn, 1995, Appleton & Lange.

MONEY-BACK GUARANTEE

We are confident that ACE THE BOARDS will prepare you for passing the USMLEs. We are so sure of this, that we'll offer you a money-back guarantee should you fail the USMLE. To receive your refund, simply mail us a copy of your failed USMLE report, plus the original receipt for this product. Mail these materials to:

Marketing Manager, medical textbooks
Mosby–Year Book, Inc.
11830 Westline Industrial Drive
St. Louis, MO 63146

Contents

PART 1

Foundations 1

1. Biology of Behavior 3
   - Genetics 3
   - Neuroanatomy 5
   - Behavioral Biochemistry 6
   - Multiple Choice Review Questions 12

2. Psychosocial Models of Behavior 13
   - Psychodynamics 13
   - Learning Theory 17
   - Cognitive Behavioral Concepts 21
   - Multiple Choice Review Questions 23

3. Measurement of Behavior 24
   - Basic Testing Concepts 24
   - Assessment of Ability and Achievement 25
   - Assessment of Personality and Psychopathology 27
   - Neuropsychologic Assessment 29
   - Epidemiology and Statistics 30
   - Multiple Choice Review Questions 35

PART 2

Development and the Life Cycle 37

4. Pregnancy and Infancy 39
   - Pregnancy 39
   - Childbirth 40
   - Neonatal and Infant Behavior 41
   - Multiple Choice Review Questions 45

5. Childhood and Adolescence 46
   - Early Childhood (1 to 6 Years) 46
   - School-Aged Children (6 to 12 Years) 48
   - Adolescence (12 to 18 Years) 48
   - Multiple Choice Review Questions 50

6. Early and Middle Adulthood 51
   - Developmental Features 51
   - Family and Relationships 51
   - Work and Leisure 52
   - Midlife Issues 53
   - Religious Commitment 54
   - Multiple Choice Review Questions 55

7. Aging, Death, and Dying 56
   - Characteristics of the Elder Population 56
   - Changes in Later Life 56
   - Problems of Later Life 58
   - Death and Dying 59
   - Multiple Choice Review Questions 62

PART 3

Behavioral Health 63

8. Doctor-Patient Relationships 65
   - Clinical Communication 65
   - Adherence and Compliance 66
   - Emotion and Behavior in Illness 67
   - Multiple Choice Review Questions 69

9. Stress 70
   - Defining Stress 70
   - Physiology of Stress 71
   - Modifiers of Stress 71
   - Stress Management 72
   - Stress and Medical Care 73
   - Chronic Pain 73
   - Multiple Choice Review Questions 75

10. Substance Use 76
    - Basic Concepts 76
    - Alcohol Abuse 76
    - Tobacco Use 78
    - Other Drugs 79
    - Multiple Choice Review Questions 83

11. Sleep 84
    - Normal Sleep 84
    - Sleep Disorders 84
    - Multiple Choice Review Questions 88

Part 4

Psychopathology 89

12 Mood Disorders 91
Basic Concepts 91
Clinical Features 92
Etiology 93
Treatment of Mood Disorders 94
Multiple Choice Review Questions 96

13 Anxiety Disorders 97
Basic Concepts 97
DSM-IV Disorders 97
Treatment of Anxiety Disorders 100
Multiple Choice Review Questions 101

14 Schizophrenia and Related Disorders 102
Basic Concepts 102
Signs and Symptoms 102
Course and Prognosis 104
Etiology 105
Treatment of Schizophrenia 105
Other Disorders 106
Multiple Choice Review Questions 107

15 Cognitive Disorders 108
Basic Concepts 108
Delirium 108
Dementia 109
Amnestic Disorder 111
Multiple Choice Review Questions 112

16 Disorders Associated with Childhood and Adolescence 113
Developmental Disorders 113
Other Disorders Associated with Childhood 115
Multiple Choice Review Questions 118

17 Other Psychiatric Disorders 119
Personality Disorders 119
Eating Disorders 119
Somatoform and Factitious Disorders 121
Dissociative Disorders 122
Sexual and Gender Identity Disorders 123
Multiple Choice Review Questions 126

Part 5

Social Context 127

18 Health Care Delivery and Economics 129
Physicians and Other Health Care Personnel 129
Health Care Delivery Systems 129
Financing Health Care 132
Multiple Choice Review Questions 135

19 Legal and Ethical Issues 136
Confidentiality 136
Consent and Informed Consent 137
Competence 137
Civil Commitment 138
Advance Directives 138
Malpractice 139
Criminal Responsibility 139
Multiple Choice Review Questions 140

20 Medicine and the Social Sciences 141
Socioeconomics of Health Care 141
Illness as a Social Phenomenon 142
Cultural Change and Diversity 143
Health Behavior 144
"Alternative" Medicine and Related Issues 145
Multiple Choice Review Questions 147

Answers and Explanations to Multiple Choice Review Questions 149
Chapter 1   149
Chapter 2   149
Chapter 3   150
Chapter 4   151
Chapter 5   151
Chapter 6   152
Chapter 7   153
Chapter 8   153
Chapter 9   154
Chapter 10 155
Chapter 11 155
Chapter 12 156
Chapter 13 156
Chapter 14 157
Chapter 15 158
Chapter 16 158
Chapter 17 159
Chapter 18 159
Chapter 19 160
Chapter 20 161

Behavioral
Science

Part 1
Foundations

Chapter 1

Biology of Behavior

GENETICS

■ *Methods* of behavioral genetics emphasize studying the relationship between family membership and the occurrence of an illness or characteristic.

- *Pedigrees* are established when one tracks occurrence of a trait or illness in a family tree.

  • A *proband* is an individual who has the illness or characteristic in question and whose relatives are examined in *family risk studies.*

  • Finding higher incidence within the family than within the general population is highly suggestive, and particular patterns of occurrence can provide additional information about modes of inheritance.

  • *Consanguinity* describes people linked by descent from a common ancestor.

- **Twin studies**

  • *Concordance* is an expression of the frequency with which an illness or trait found in one twin is found in the other.

  • *Monozygotic (identical) twins* are of interest because they represent a dyad of maximum genetic similarity.

  • *Dizygotic (fraternal) twins* are less similar genetically because they represent fertilization of two separate ova.

  • Concordance rates among monozygotic (MZ) and dizygotic (DZ) twins are often compared with the expectation that a genetic basis for the trait will be reflected in a higher concordance rate among the genetically identical MZ twins.

- **Adoption studies**

  • The study of children adopted and thus reared apart from those with whom they share genetic material helps to control for environment and parenting in the development of a behavioral characteristic or disorder.

  • One strategy involves examining children of affected parents reared by adoptive parents who are not affected.

  • Another strategy involves studying twins reared separately.

- *Heritability* is a numerical estimate of the relative importance of genetic factors in determining whether an individual will possess a characteristic or disorder.

- The study of genetic factors in behavior is rendered much more complex by the apparent *multifactorial* nature of inheritance in most characteristics and disorders.

Table 1.1	Family Correlations for IQ Values
RELATIONSHIP	MEDIAN CORRELATION
Monozygotic twins (reared together)	0.85
Dizygotic twins (reared together)	0.58
Siblings (reared together)	0.45
Cousins	0.15

From Bouchard TJ Jr, McGue M: Familial studies of intelligence: a review, *Science* 212:1055, 1991.

Table 1.2	Genetic Defects Associated with Retardation
DISORDER	GENETIC DEFECT
Down syndrome	Abnormality in chromosome 21
Klinefelter's syndrome	Multiple X chromosomes
Fragile X syndrome	Mutation on the X chromosome
Turner's syndrome	XO pattern in sex chromosomes

Influences on Behavior

- *Temperament* refers to behavioral features evident in infants from birth.
    - Relevant characteristics include activity level, nature and intensity of reaction to stimuli, mood, distractibility, and persistence.
    - Concordance in temperament characteristics is higher in MZ twins than in DZ twins.

- **Intellect**
    - Familial studies clearly show higher concordance in IQ scores in dyads with higher presumed genetic similarity (Table 1.1).
    - Some forms of mental retardation have known genetic causes (Table 1.2).

- **Personality characteristics**
    - MZ twins show higher concordance in personality characteristics than DZ twins do.
    - MZ twins reared together and apart show similar levels of concordance in personality characteristics.
    - Familial studies indicate genetic factors in a variety of personality disorders, including borderline, antisocial, schizoid, histrionic, and others.
    - Some personality disorders are associated with higher than normal incidence of specific psychiatric problems in relatives of affected patients. For example, *antisocial personality disorder* is associated with *alcoholism, learning disorders,* and *attention deficit disorder* among relatives.

- **Psychopathology**
    - Genetic factors have been implicated in a broad range of psychiatric disorders (Box 1.1).

Chapter 1 Biology of Behavior

> **Box 1.1**
>
> | **MAJOR DISORDERS ASSOCIATED WITH GENETIC FACTORS** |
> | --- |
> | Alcoholism |
> | Anxiety disorders (panic disorder, agoraphobia) |
> | Mood disorders (depression, bipolar disorder) |
> | Schizophrenia |
> | Somatization disorder |
> | Anorexia |
> | Alzheimer's disease |
> | Gilles de la Tourette's syndrome |
> | Pervasive developmental disorder (autism) |

- The risk of these disorders in first-degree relatives of index patients is often several times higher than that found in the general population.
- Concordance rates are found to rise with increasing genetic similarity.

NEUROANATOMY

Structural Anatomy

- *Neurons* are the fundamental functional units of the nervous system.
    - *Axons* are the processes that carry information to other neurons.
    - Efficiency of axonal conduction may be enhanced by a coating called a *myelin sheath.* Damage to the sheath (e.g., from multiple sclerosis) can produce a broad range of neurologic dysfunction.
    - *Axon terminals* incorporate *synaptic vesicles* containing neurotransmitters.
    - *Dendrites* are the branching processes on which are found receptor sites for messages projected across the synaptic cleft from the axons of other neurons.
- *Glia* include several different types of nonneuronal cells in the central nervous system (CNS) and peripheral nervous system (PNS).
    - Glial cells outnumber neurons in the nervous system by a factor of 10:1.
    - Glia provide physical structure and some (e.g., astrocytes) are part of the blood-brain barrier.
    - Other glia (oligodendroglia and Schwann cells) provide the myelin sheath.
- Systems
    - The *CNS* includes the brain and the spinal cord.
    - The *PNS* includes the sensory and motor pathways to and from the CNS, the cranial nerves, and other structures outside the CNS.
        - The *autonomic nervous system* innervates the internal organs.
        - The *sympathetic division* activates organs to facilitate activity and energy expenditure.

- The *parasympathetic division* largely functions to inhibit the effects of sympathetic innervation.

■ **Behavioral Geography**
- The *cerebral cortex* consists of the two cerebral hemispheres, each divided anatomically into *frontal, temporal, parietal,* and *occipital* lobes.
  - The anatomic divisions are associated with some functional specializations (Table 1.3).
  - The cerebral hemispheres are joined by several structures called *commissures,* the largest of which is the *corpus callosum.* Functioning in the hemispheres is *lateralized.*
    - The *left* hemisphere is specialized for language in virtually all right-handed persons and the majority of left-handed persons.
    - The *right* hemisphere is similarly specialized for spatial and visuospatial skills and nonverbal aspects of language.
    - Cortical connections with motor and sensory functioning in the body are *contralaterally organized.* For example, injury to the left frontal area might cause right-sided motor impairment and expressive speech deficits.
  - *Cytoarchitectonic* division of the cortex is a feature that allows one to identify areas grouped together in terms of similar cell types and structures.
- **Subcortical structures**
  - The *limbic system* consists of a group of structures important in affect, motivation, and appetitive behavior (Table 1.4).
  - Limbic system structures also play important roles in memory functioning.
  - The *reticular activating system* includes structures involved in bodily homeostasis as well as regulating alertness and level of consciousness.
  - The *cerebellum* controls coordination and organization of body movement.
- *Lesion-behavior correlations:* Lesions in the nervous system are often associated with characteristic behavioral effects (see Table 1.4).

BEHAVIORAL BIOCHEMISTRY

■ *Neurotransmitters* are the chemical substances that carry out communication between neurons.

Table 1.3 Cortical Lobes and Associated Functions

Lobe	Associated Functions
Frontal	Motor functions, higher order reasoning, judgment
Temporal	Auditory processing, memory
Parietal	Somatosensory processing
Occipital	Visual processing

- Neurotransmission

    • Transmitters released from *vesicles* in the axon terminal of the *presynaptic* neuron cross the *synaptic cleft* and are taken up by *receptors* in dendrites of the *postsynaptic* neuron.

    • The influence of the transmitter on the postsynaptic neuron may be *excitatory* or *inhibitory*. Neurons may have receptors for more than one neurotransmitter.

    • Pharmacologic manipulation of neurotransmitter activity may take several strategies:

    - *Increasing the plasma levels of precursors*
      Example: Administration of L-dopa to patients with Parkinson's disease
    - *Receptor blockade*
      Example: Antipsychotic medications block dopamine receptors.
    - *Inhibition of reuptake*
      Example: Antidepressants, such as fluoxetine, inhibit reuptake of serotonin.
    - *Inhibition of oxidation*
      Example: Antidepressants known as monoamine oxidase inhibitors prevent breakdown of norepinephrine and serotonin.

Table 1.4 *Behavioral Correlates of Brain Lesions*

Structure	Lesion Effects
Cortex	
Frontal lobe	Contralateral motor deficits
	Impairment in judgment and social behavior, personality change
	Deficits in attention and motivation
Temporal lobe	Memory impairment
	(*Left*) Impaired receptive language
	(*Right*) Impaired nonverbal auditory perception (e.g., tone, rhythm)
Parietal lobe	Contralateral impairment in somatosensory perception (neglect, agnosia)
	(*Left*) Impaired perception of verbal symbols (numbers, letters)
	(*Right*) Impaired constructional skills
Occipital lobe	Contralateral visual impairments
Subcortical	
Hypothalamus	Altered eating and sexual behavior
	Sleep disturbances
Hippocampus	Memory impairment
Amygdala	Uncontrolled rage or excessive docility
	Hypersexuality
Thalamus	Altered pain perception, memory deficit
Basal ganglia	Movement disorder (Parkinson's disease)
	Mood disorders
Reticular formation	Disturbances in sleep and level of consciousness
Cerebellum	Ataxia, impaired position sense

- Major classes of neurotransmitters include the *biogenic amines* (or monoamines), *amino acids,* and *neuropeptides.*
- *Biogenic amines* are referred to as functionally specific neurotransmitters because they are found in more restricted locations and associated with more specific functions.

  — ***Dopamine***
    Class: catecholamine
    Behavioral associations
    - Schizophrenia
    - Movement-related disorders, such as Parkinson's disease and Gilles de la Tourette's syndrome

    Synthesis: converted from tyrosine by the action of tyrosine hydroxylase
    Areas of concentration
    - *Nigrostriatal tract:* regulation of muscle tone and movement
    - *Tuberoinfundibular tract:* regulation of prolactin secretion from the anterior pituitary
    - *Mesolimbic—mesocortical tract:* modulation of mood and appetitive behavior

    Applications

    - Medications that enhance availability of dopamine are used in treatment of Parkinson's disease and related disorders.
    - Medications that block dopamine receptors are used in treating schizophrenia.

  — ***Norepinephrine***
    Class: catecholamine
    Behavioral associations: regulation of arousal, sleep, pain/pleasure, learning, and mood
    Synthesis: synthesized from dopamine by beta-hydroxylase
    Area of concentration: locus ceruleus
    Applications

    - Medications that enhance availability of norepinephrine are used as antidepressants and also have uses in treatment of sleep disturbances and chronic pain.

  — ***Serotonin***
    Class: indolamine
    Functions: regulation of mood, sleep, and pain
    Synthesis: converted from tryptophan by tryptophan hydroxylase
    Area of concentration: raphe nuclei
    Applications

    - Medications that inhibit serotonin reuptake or degradation by monoamine oxidase are employed as antidepressants and are used in the treatment of chronic pain, headaches, sleep disturbances, and eating disorders.

  — ***Acetylcholine (ACh)***
    Class: quaternary amine

Functions

- Acetylcholine is implicated in motor functions by virtue of its role in nerve-skeleton-muscle junctions.
- It is implicated in learning and memory by virtue of association with Alzheimer's disease.

Synthesis: catalyzed from acetyl coenzyme A and free choline by choline acetyltransferase

Areas of concentration

- The septum, which supplies ACh to the septal-hippocampal tract
- The nucleus basalis of Meynert

Application

- *Tacrine,* an acetylcholinesterase inhibitor, has been approved for treatment of Alzheimer's disease, though only mild benefit has been reported.
- Anticholinergic medications, such as *benztropine* (Cogentin), which block the *muscarinic* class of ACh receptors, are used in the treatment of medication-induced movement disorders such as those observed in patients treated with antipsychotics.
- Anticholinergic medications have also been used in treatment of idiopathic Parkinson's disease.

— **Histamine**

Class: ethylamine

Functions: implicated in movement, sleep, and anxiety

Area of concentration: hypothalamus

Applications

- Antihistamines are common ingredients in over-the-counter hypnotics.
- Antihistamines, such as hydroxyzine and diphenhydramine, are employed as short-term anxiolytics and as antiparkinsonian agents.
- *Cyproheptadine* is used in treatment of anorexia because it may promote weight gain and in the treatment of inhibited orgasm secondary to serotonergic medications.

- *Amino acid transmitters* are referred to as *global transmitters* because they are much more widely distributed in the nervous system. They are found in approximately 70% of synapses.

    - The amino acid transmitters include *glutamate, gamma-aminobutyric acid (GABA), glycine,* and *aspartate.* Their influence and locations are given in Table 1.5.
    - GABA hypoactivity is implicated in *anxiety disorders.*
        - The *benzodiazepine* anxiolytics act by increasing the affinity of GABA for its binding sites, and barbiturates are also believed to exert anxiolytic effects through GABA receptors.
    - Both GABA and glutamate are implicated in the development of seizure disorders.

Table 1.5 Amino Acid Transmitters

Transmitter	Influence	Distribution
GABA	Inhibitory	Brain, dorsal spinal cord
Glutamate	Excitatory	Brain, dorsal spinal cord
Glycine	Inhibitory	Ventral spinal cord
Aspartate	Excitatory	Ventral spinal cord

GABA, Gamma-aminobutyric acid.

Table 1.6 Psychiatric Disorders and Related Neuropeptides

Disorder	Neuropeptides
Schizophrenia	Cholecystokinin, neurotensin
Mood disorders	Vasopressin, substance P
Dementia of the Alzheimer's type	Somastatin, substance P
Huntington's disease	Somastatin, substance P
Eating disorders	Cholecystokinin

Table 1.7 Endocrine Disorders and Psychopathology

Endocrine Disorder	Psychopathology
Hyperthyroidism	Anxiety, agitated depression, psychosis, manic excitement, memory impairment
Hypothyroidism	Depression, paranoia
Hyperparathyroidism	Personality changes, apathy, delirium, cognitive impairment
Hypoparathyroidism	Personality changes, delirium
Adrenocortical insufficiency	Irritability, depression, apathy, psychotic reactions
Adrenocortical hyperproduction	Agitated depression, memory impairment, attention deficits, occasional psychotic reactions

- Both *benzodiazepines* (e.g., diazepam, Valium) and *barbiturates* (e.g., phenobarbitol) have applications as anticonvulsants.

• *Neuropeptides* are proteins constituted of less than 100 amino acids. Several dozen have been identified, but many are as yet poorly understood.

- The peptides are synthesized on ribosomes in the cell body as portions of protein precursors called *proproteins.*
  - Proproteins are hydrolyzed to produce active peptides during transport to the terminal.
- The *endogenous opioids* include the peptides referred to as *enkephalins* and *endorphins,* which bind to morphine receptors.
  - These peptides are implicated in stress, pain, and mood.

- Some neuropeptides are believed to be associated with specific psychiatric disorders (Table 1.6).
- The peptides include the *modulatory transmitters* whose action is not the activation or cessation of functions but modification of the processes. Many of these neuropeptides are released onto the presynaptic terminal of another synapse, diminishing or augmenting the action of that synapse.

■ **Hormones and Behavior** A variety of psychiatric symptoms and syndromes are associated with endocrine abnormalities (Table 1.7).

MULTIPLE CHOICE REVIEW QUESTIONS

1. A familial form of a disorder or characteristic is identified by studying which of the following?
   a. Concordance
   b. Consanguinity
   c. Pedigree
   d. Heritability

2. A stroke patient speaks correctly but without animation or inflection, and does not appreciate what others communicate in their tone of voice. The damage from the stroke most likely involves which of the following?
   a. Right cerebral hemisphere
   b. Left cerebral hemisphere
   c. Basal ganglia
   d. Hippocampus

3. After a head injury, a patient shows dramatic personality change, becoming impulsive, profane, and sexually inappropriate. Lesions are particularly likely to involve which of the following?
   a. Thalamus
   b. Parietal lobe
   c. Reticular formation
   d. Frontal lobes

4. The neurotransmitter most specifically associated with movement disorders such as Parkinson's disease and Tourette's syndrome is which of the following?
   a. Norepinephrine
   b. Serotonin
   c. Dopamine
   d. Gamma-aminobutyric acid

5. Medications that will be of use in the treatment of Alzheimer's disease are most likely to do which of the following?
   a. Increase availability of acetylcholine
   b. Block dopamine receptors
   c. Inhibit serotonin reuptake
   d. Inhibit the activity of monoamine oxidase

Chapter 2
Psychosocial Models of Behavior

PSYCHODYNAMICS

- The essence of psychodynamic thinking is the idea that behavior is the outcome of the interplay of conflicting internal forces, both biologic and psychologic in their origin.
- Many of the concepts in psychodynamic thinking have their origins in the psychoanalytic theory of Sigmund Freud, but many of these concepts have acquired broader meaning and more general application in modern theory and practice.

Principles

- The idea of *unconscious motivation* is that awareness of the influences on behavior (i.e., needs, drives, wishes) is incomplete.
- *Conflict* is generally considered inevitable as the organism seeks to satisfy competing and sometimes incompatible drives and needs.
- *Drives* are impulses for behavior that reflect fundamental biologic forces and *instinctual* elements in behavior.
    - *Sexuality* (also referred to as *libido*) and *aggression* are considered to be especially prominent areas of drive-related behavior.
    - Drives are believed to reflect two broad types of instinctual influence:
        - *Eros:* the life-affirming instincts reflected in self-preservation and species preservation by sexual propagation.
        - *Thanatos:* the so-called *death instinct* expressed in violence and self-destructive behavior.
    - Unsatisfied drives impel behavior by creating a state of tension that the organism will seek to reduce. *Drive reduction* is discharge of that tension.

Levels of Consciousness

Mental life takes place at different levels of awareness. This division of mental life is incorporated in the *topographic theory of the mind.*

- *The unconscious mind* incorporates all those aspects of mental life that lie outside of awareness and are not readily accessible.
    - The major part of mental life is believed to exist outside of consciousness.
    - Unconscious motives and thoughts are most often accessible only under special circumstances. Therapeutic techniques, such as *free association* and *interpretation,* are designed to help bring unconscious material into awareness.

- *Dreams* are believed to reflect unconscious thoughts and desires, though not in a direct or logical way.
- Unacceptable thoughts and motives are relegated to the unconscious mind by the mechanism of *repression.* The threat that such material may emerge into consciousness is an important source of *anxiety.*

● *The preconscious mind* incorporates those aspects of mental life that lie outside of awareness at any given time but may be readily accessed.

- *Example:* You may not be thinking at this moment about where you went to high school, but you could readily retrieve the facts from memory.

● *The conscious mind* incorporates those aspects of mental life that lie within immediate awareness.

Structure of Personality

The topographic theory of the mind is complemented by a *structural theory of the mind* in which personality is divided into the three components *id, ego,* and *superego.*

● **Id**

- The id is the repository of the biologically based and instinctual drives and impulses.
- The id operates on the *pleasure principle,* in which behavior is governed by the sole factor of minimizing discomfort and pain and maximizing pleasurable satisfaction of drives.
- Mental life at the level of the id is dominated by *primary-process thinking.*
   - Primary process thinking comprises drives, wishes, fantasies, and desires ungoverned by considerations of logic or reality.
   - The influences of the id are primarily unconscious, and unconscious mental life is dominated by primary process thinking.
- The id is present from birth and represents primitive and undeveloped mental life. It is reflected in the *magical thinking* and *wish fulfillment* observed in children.

● **Ego**

- The ego is the rational and executive agency of personality.
- The ego operates on the *reality principle.*
   - Drives are regulated and gratification postponed until appropriate circumstances are achieved.
   - The ego develops from the id to facilitate pursuit of needs in a manner consistent with the requirements of the real world.
- Mental life at the level of the ego is dominated by *secondary-process thinking,* emphasizing problem solving and realistic appraisal of the world.
   - The ego is the agency of *reality testing,* or the process of deciding what is valid about the outside world and the self.
   - Ego functions include all the mental processes incorporated in the concept of *cognition,* including perception, learning, attention, memory, language, and reasoning.
   - Ego functions also include interpersonal relationships and the *de-*

Chapter 2   Psychosocial Models of Behavior

fense mechanisms, which are strategies for managing conflict and distress (see the discussion of defense mechanisms, p. 16).

- **Superego**
  - The *superego* is the internal representation of the values and standards learned from family and culture. It includes the *conscience* and the *ego ideal.*
  - The conscience administers internal rewards and punishments for approved and disapproved behavior.
  - The ego ideal is the individual's mental model of moral and behavioral perfection.
  - Like the id and unlike the ego, the superego is a nonrational agency of personality. No formal principle is assigned to the superego, but in effect its governing principle is the pursuit of perfection.

Psychosexual Development

- *Freud's stages* divided development into periods reflecting association of pleasure with different *erogenous zones.*
  - "Sexuality" in childhood referred generally to experiences of pleasure from stimulation of the body.
  - The earliest stages (*oral* and *anal*) gave primacy to the physical experiences of feeding and elimination and the interpersonal experiences of feeding and toilet training.
  - The *phallic* stage features the *oedipal* conflict in which genital experiences are primary. The child must deal with erotic impulses toward the opposite-sex parent, which imply competition with the same-sex parent.
  - In the *latency* stage, drive activity subsides until adolescence and the emergence of the *genital* stage, when truly adult sexuality may develop (Table 2.1).

- *Erikson's stages* emphasized interpersonal experiences and *psychosocial* development and extended development throughout the life-span.
  - Each stage involves a characteristic developmental challenge.
  - *Example:* The experience of toilet training at 2 years of age is important not because of instinctual drives related to anal pleasures but because

Table 2.1   *Freud's Stages of Psychosexual Development*

STAGE	AGE IN YEARS	FEATURES
Oral	0-1	Focus on feeding, sucking, biting, and tasting as primary sources of pleasure
Anal	2-3	Focus on physical experiences associated with elimination and control of the sphincters; also includes toilet training
Phallic	4-6	Focus on the genitalia as source of pleasure; children may masturbate and display interest in genitals of others; includes the **oedipal conflict**
Latency	6-12	Identification with the same-sex parent
Genital	12+	Learning to express sexuality in the context of mutual adult relationships

Table 2.2	Erikson's Stages of Psychosocial Development
AGE	DEVELOPMENTAL CHALLENGE
0-18 months	Basic trust versus mistrust
18-36 months	Autonomy versus shame and doubt
3-6 years	Initiative versus guilt
6-11 years	Industry versus inferiority
Adolescence	Identity versus role diffusion
Early adult	Intimacy versus isolation
Middle adult	Generativity versus stagnation
Old age	Integrity versus despair

of the challenge of learning *autonomy* (control over the body) versus experiencing *shame and doubt* from failure.

- Adult stages emphasize the capacity for *intimacy*, contributing to the development of others *(generativity)* and coping with the realities of aging *(ego integrity)* (Table 2.2).

- *Identification* is the process of incorporating the characteristics and qualities of another (e.g., parents) into personality.

- *Fixation* occurs when the tasks of a stage in development are not achieved. This has adverse effects on personality development and on coping with later developmental challenges.

■ Defense mechanisms are strategies for dealing with feelings, impulses, and desires that would cause conflict, anxiety, or distress if admitted fully to consciousness (Table 2.3).

■ Psychoanalytic therapy emphasizes achievement of *insight* or a fuller awareness of unconscious motivations.

- Classical psychoanalysis usually involves four or more treatment sessions over a period of 2 to 5 years.
  - Emphasis is on bringing *repressed* impulses and feelings into awareness.
  - Classical techniques include *free association, interpretation, dream work,* and *analysis of transference* (Table 2.4).

- *Psychoanalytic psychotherapy* is more commonly practiced today than classical analysis.
  - In addition to the techniques of classical analysis, the therapist employs more direct discussion of material.
  - The therapist may play a more active role in discussion of material and may be more involved in supporting the integrity of the patient's ego functions.

- *Transference and countertransference*
  - *Transference* occurs when the patient's relationship with the therapist is influenced by unconscious feelings about figures from the past (e.g., parents).
  - *Countertransference* occurs when the way the therapist relates to the patient is influenced by feelings about figures from the past.

Table 2.3 *Defense Mechanisms*

Mechanism	Characteristics
Repression	Unconscious exclusion of external distressing material from awareness
Projection	Attribution of undesirable thoughts and impulses to others
Regression	Reversion to more immature patterns of behavior under stress
Identification	Adoption of the characteristics and behaviors of others
Reaction formation	Adopting attitudes and behaviors opposite to unconsciously held but unacceptable or threatening feelings
Intellectualization	Avoidance of emotion and focusing exclusively on the abstract and intellectual aspects of mental life
Rationalization	Generating plausible justifications for irrational or threatening ideas
Displacement	Redirecting emotions from an original object to a more acceptable or available substitute
Denial	Unconscious refusal to recognize unacceptable realities
Splitting	Restricting emotional reactions to only one side of feelings about conflicted issues
Sublimation	Redirecting instinctual energies and drives into more adaptive activities
Suppression	Consciously directing attention away from internal unwanted feelings and impulses

Table 2.4 *Psychoanalytic Techniques*

Technique	Features
Free association	Patient is to express whatever comes to mind without censorship, to encourage emergence of unconscious material
Dream work	Exploring dreams for *latent* meanings about unconscious feelings
Analysis of the transference	Learning about unconscious feelings about past relationships from the way those feelings influence relationship with the therapist
Interpretation	Identifying associations between motives and behavior, the significance and meaning of uncovered material, the evidence of unrecognized motives, and so on

LEARNING THEORY

- *Learning* is defined as a relatively enduring change in behavior as a result of experience.
- *Learning theory* is an effort to formulate the rules by which behavior is acquired and changed.
- Two major learning models involve *classical* and *operant conditioning*.

■ Classical Conditioning

The essence of classical conditioning is the association of naturally occurring ("reflexive") behavior with new stimuli that do not ordinarily elicit the behavior. The process was first identified and studied by the Russian physiologist *Ivan Pavlov*.

- **Stimuli and responses**
    - An *unconditioned stimulus* automatically elicits reflexive behavior without experience or training. The behavior it elicits is called an *unconditioned response.*
    Example: A puff of air in the eye (stimulus) elicits blinking (response).
    - During *acquisition,* the organism learns to give that response to a previously neutral stimulus.
    Example: Each time there is a puff of air, a bell sounds. After several trials, the bell alone will elicit blinking.
    - When the behavior is elicited by the new stimulus, it is called a *conditioned response,* and the new stimulus is called a *conditioned stimulus.*

- **Extinction and spontaneous recovery**
    - *Extinction* occurs when the association between the conditioned and unconditioned stimuli is no longer reinforced. The conditioned stimulus eventually will no longer elicit the response.
    Example: The bell is repeatedly presented without the puff of air. Soon, the bell no longer elicits blinking.
    - In *spontaneous recovery* the response to the conditioned stimulus returns, even though extinction has taken place.
    Example: Some time later the bell is presented alone and blinking again occurs.

- **Generalization and discrimination**
    - *Generalization* occurs when the response is elicited not only by the original stimulus, but also by similar or related stimuli.
    Example: A bell with a different sound is presented and still elicits blinking.
    - In *discrimination* the organism learns to respond more specifically.
    Example: Only one bell tone is accompanied by a puff of air, whereas other tones are never paired with a puff. Soon blinking occurs only to the one tone.

- *Aversive conditioning* involves pairing stimuli with painful or unpleasant responses.
    - Aversive techniques are sometimes applied to the treatment of addictions and other undesirable behavior.
    Example: A pedophile receives a painful electric shock each time he responds with arousal to pictures of children, and the arousal response diminishes.
    - Aversive conditioning is also subject to the phenomena of extinction, discrimination, and so on; these phenomena limit the utility of aversive techniques.

Operant Conditioning

The essence of operant conditioning is learning connections between behaviors and their consequences. The study of operant conditioning is most strongly associated with the work of the American psychologist *B.F. Skinner.*

- **Operant versus classical conditioning**
    - Operant conditioning is not limited to reflexive or automatic behaviors and thus helps to account for the wide variety of behaviors that humans can learn.
    - The role of the *stimulus* is not to *elicit* a specific response. Stimuli influence the probability that the organism will *emit* various behaviors in its *repertoire.*

Chapter 2 Psychosocial Models of Behavior   19

Example: The internal state of hunger and the external event of seeing a sign for your favorite restaurant increase the likelihood that you will enter and order food but do not elicit this automatically or reflexively.

- The concepts of *acquisition, extinction, generalization,* and *discrimination* are relevant to both kinds of conditioning.

● *Reinforcement* refers to consequences of behavior that *increase* the likelihood that the behavior will be performed again.

- *Positive reinforcement* occurs when a desirable or positive consequence increases the behavior that preceded it.

Example: Praise for his playing increases a child's practice at the piano.

- *Negative reinforcement* occurs when the removal of an undesirable or aversive stimulus increases the behavior that preceded it.

Example: A dawdling child pays more attention to completing his classwork to avoid being left alone when the others are finished and have time to play outside.

● *Punishment* refers to consequences that *decrease* the likelihood a behavior will occur again.

- It may involve the *presentation* of an aversive stimulus (e.g., putting a child in time-out) or *withdrawal* of a desirable one (e.g., ordering the child out of the swimming pool).
- The effectiveness of punishment in producing behavior change is limited by several factors:
    - It may simply suppress the behavior in limited situations.
    - It tends to provoke aggressive responses and is more likely to breed hostility.
    - It does not in itself increase the likelihood of desirable behavior.

● **Schedules of reinforcement**

- The effects of reinforcement are influenced by the frequency and timing of reinforcers. The patterns in which reinforcements occur are called *schedules of reinforcement.*
- Reinforcers may depend on the frequency of a behavior (*ratio* schedules) or the passage of time (*interval* schedules). The ratio or interval may be *fixed* or *variable* (Table 2.5).
- In general, behavior that is *intermittently* reinforced will be harder to extinguish than behavior that is *continuously* reinforced.

● *Shaping* involves progressive refinements of behavior through experience and is important in understanding how complex behaviors are learned.

- *Successive approximation* refers to gradually requiring more and more precise versions of the desired behavior in order for reinforcement to occur.
- The idea of shaping also supplies to gradual modification of the situations in which a behavior occurs.

Example: Shaping a child's social behavior involves teaching the child not only what to do but also when to do it.

■ **Applications**

Behavioral techniques have applications in the treatment of psychiatric disorders, management of child-rearing problems, elimination of unhealthy habits, and treatment of medical problems, such as obesity and headache.

Table 2.5 Schedules of Reinforcement

Schedule	Reinforcement	Example
Continuous	After each behavior	A cookie for each patient you examine
Fixed ratio	After a fixed number of responses	A cookie for every fifth patient you examine
Variable ratio	After an average but varying number of responses	Cookies on average for every fifth patient, but sometimes after the third or the eighth or some other number
Fixed interval	After a fixed period of time	Cookies every Friday no matter how many patients were examined
Variable interval	After randomly varying periods of time	Cookies may come on any day whether you saw any patients

- *Exposure therapies* are used to treat *phobias* through controlled exposure to the fear-provoking stimulus.

  - In vivo exposure takes place in the actual fear-provoking situation (e.g., riding an elevator).
  - *Systematic desensitization* is a specific technique conducted in the office or laboratory.
    - Progressively stronger versions of the fear-provoking stimulus are presented.
    - The subject is taught a relaxation technique and moves up to stronger stimuli only when relaxation can be maintained for weaker stimuli.
  - Exposure therapies, such as desensitization, are based on classical conditioning. The approach interrupts the automatic nature of the fear response.
  - *Flooding* is an exposure technique that is not gradual but instead presents the fear-provoking stimulus at full strength while the patient is assisted in using coping techniques.

- *Biofeedback* techniques are based on the observation that control over usually involuntary physical functions can be exerted if there is information available about the status of the function.
Example: Patients can learn to control the tension in the frontalis muscle in the forehead when electromyographic data are converted into an auditory or visual signal.

  - Biofeedback has been applied to the treatment of anxiety, tension and migraine headaches, and hypertension.
  - Apart from the technical requirements of feedback about the physical function, the techniques also require *motivation* in that considerable practice is necessary.

- *Behavior modification* involves application of operant conditioning principles to alter behavior.

  - *Contingency management* is the general term for manipulating the circumstances under which reinforcement occurs.

Chapter 2   Psychosocial Models of Behavior

- *Token economies* are a specific and usually highly organized application of contingency management.
  - Most often they are used in group settings, such as psychiatric wards, programs for the mentally retarded, and behavior disorder classrooms.
  - Instead of immediate physical rewards, subjects earn *tokens* (or stars, or tickets) that are required for later rewards (e.g., you must have three tokens to see the movie that night).
  - Subjects earn tokens for desired behavior (washing without reminder, showing up on time for group therapy) and may lose tokens for undesirable behavior.

COGNITIVE BEHAVIORAL CONCEPTS

Cognitive behavioral models of behavior attempt to integrate behavioral concepts and techniques with the presumed influences of ideas, expectations, and other "internal" mental processes.

- *Modeling and observational learning* involve change in behavior as a result of exposure to the behavior of others.
  - Classic studies in modeling have shown how aggressiveness in children is modified by their exposure to aggressiveness modeled by others.
  - Modeling has been beneficially employed in modification of fear (e.g., dental phobias) and teaching social skills.
- *Expectancies* are beliefs about the likely consequences of events and behaviors.
  - Expectancies are implicated in the practice of health-related behaviors, such as checkups and exercise; practice depends on the belief that the behavior will have a desired effect.
  - *Self-efficacy expectancies* are generalized expectations about one's ability to carry out behaviors in a way that leads to the desired outcome.
  - *Locus of control* is an extensively studied generalized expectancy about what controls the likelihood of desirable and undesirable events.
    - *Internal locus of control* refers to the belief that, in general, reinforcement depends on one's own actions.
    - *External locus of control* refers to the belief that, in general, reinforcement depends on luck or powerful others.
    - Either position can have benefits or create problems for health care.
      - An internal locus may promote taking more responsibility but may mean less willingness to follow the physician's instructions.
      - An external locus may promote more compliance but may mean more passivity.
- **Maladaptive Thinking**
  - Cognitive therapies are based on the idea that flawed assumptions and thinking contribute to personal distress and maladaptive behavior.
    - Therapy involves identifying maladaptive ideas and assumptions, clarifying how they are mistaken or counterproductive, and learning more functional alternatives.

Example: A patient has difficulty standing up for himself because he believes people will dislike him. He is taught to think about standing up for himself in a way that supports the behavior.

- Cognitive therapy has been applied with particular value to the treatment of anxiety and depression. An important figure in the cognitive treatment of depression is the psychiatrist *Aaron Beck.*
- One approach to cognitive therapy involves modification of the "internal dialog," or *self-talk.*

Example: A patient who is afraid to speak before a group is taught to interrupt the train of thought about how scared he is and how everyone will be able to see his knees knocking and replace these thoughts with more encouraging and useful ones.

Multiple Choice Review Questions

1. In psychodynamic models of behavior, a person's ability to make reasoned judgments about events and about the consequences of behavior depends on which of the following?
   a. Primary process thinking
   b. The pleasure principle
   c. Intact ego functioning
   d. The development of the ego ideal

2. A medical student finds himself irritated with a patient who reminds him of his mother, who is a melodramatic and hypochondriacal person. The student's reaction is an illustration of which of the following?
   a. The oedipal conflict
   b. Transference
   c. Identification
   d. Fixation

3. A patient undergoes chemotherapy, which induces nausea. After a few sessions the patient begins to report nausea developing on arrival at the clinic, before treatment. This is an illustration of which of the following?
   a. Stimulus generalization
   b. Spontaneous recovery
   c. Aversive conditioning
   d. Classical conditioning

4. A patient is treated for tension headaches by training with a machine that helps her learn control over muscle tension by giving a signal indicating how tense the muscles are. This illustrates which of the following?
   a. Biofeedback
   b. Aversive conditioning
   c. Systematic desensitization
   d. Flooding

5. A patient who has an internal locus of control is most likely to do which of the following?
   a. Take a passive and compliant attitude toward health care
   b. Believe that he or she cannot effectively carry out health care behaviors
   c. Believe that health is mostly a matter of luck
   d. Be an active but perhaps independent-minded participant in health care

Chapter 3

Measurement of Behavior

As with all other phenomena of interest in medicine, the value of research and clinical practice with human behavior depends on effective and appropriate measurement. Many important behaviors and behavioral characteristics are formally measured or assessed through *psychologic tests*.

In clinical practice, there is an important distinction between *testing* (the use of specific tests and measures) and *assessment* (a broader concept in which testing is integrated with history, observation, and interview).

BASIC TESTING CONCEPTS

- *Reliability and validity* are fundamental concepts in the evaluation of tests.
  - *Reliability* deals with the ability of a test to generate *reproducible* results, either when administered more than once or when used by more than one examiner.
    - *Test-retest reliability:* the ability of a test to produce the same results over repeated administration.
    - *Interrater reliability:* the extent to which different examiners will get the same results.
  - *Validity* deals with whether a test measures what it is supposed to measure. Validity is more specifically conceptualized in different ways (Table 3.1).
  - **Relationship between validity and reliability**
    - A test can be reliable without being valid.
    - Reliability, however, sets the upper boundary on validity; a test cannot be more valid than reliable.
- *Standardization* refers to consistency in test procedures, content, and interpretation.
  - In a standardized test, examiners use specified materials and procedures, record and score responses in a specified fashion, and follow specified rules for interpreting the results.
  - Standardization is crucial to comparing test results from different individuals.
  - It permits development of *norms,* or standards by which to judge results as normal or abnormal. Norms are often adjusted for characteristics such as age, sex, and education.
- **Sensitivity and Specificity**
  - *Sensitivity* is the proportion of patients actually found to have a condition when the test results say they do.

Table 3.1 Types of Validity

Type	Definition
Face	Does the test appear to measure relevant things?
Predictive	Does the test correctly identify outcomes, or future status, on related behaviors and characteristics?
Concurrent	Do test results accurately identify current status on related behaviors and characteristics?
Construct	Is the test actually a measure of the theoretical concept involved (i.e., intelligence)?

Table 3.2 Calculation of Sensitivity, Specificity, and Predictive Value

Term	Calculation
Sensitivity	$\dfrac{\text{True positives}}{\text{True positives + False negatives}}$
Specificity	$\dfrac{\text{True negatives}}{\text{True negatives + False positives}}$
Positive predictive value	$\dfrac{\text{True positives}}{\text{True positives + False positives}}$
Negative predictive value	$\dfrac{\text{True negatives}}{\text{True negatives + False negatives}}$

- *Specificity* is the proportion of patients who in fact do not have a condition when the test results say they do not.

Predictive Value

- *Positive predictive value* is the probability that positive test results mean the patient has the disease.
- *Negative predictive value* is the probability that the patient does not have the disease when the results are negative (Table 3.2).

ASSESSMENT OF ABILITY AND ACHIEVEMENT

Intelligence Tests

- *Intelligence* is a multidimensional concept referring to the abilities involved in reasoning, learning, problem solving, and manipulation of symbols.
- *Intelligence tests* are standardized procedures for sampling behaviors believed to reflect these abilities.
- The IQ score
    - Results are typically expressed as *IQ (intelligence quotient)* scores, which have a mean of 100 and a standard deviation of 15.
        - This is referred to as a *deviation IQ*.

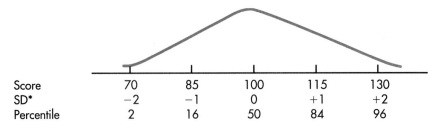

*Standard deviations

Fig. 3.1 Distribution of IQ scores

Table 3.3 Wechsler Intelligence Tests	
TEST	AGE GROUPS
Wechsler Adult Intelligence Scale—Revised (WAIS-R)	16-75 yr
Wechsler Intelligence Scale For Children—Third Edition (WISC-III)	6-16 yr
Wechsler Preschool and Primary Scale of Intelligence (WPPSI)	4-6½ yr

- Scores reflect a person's standing in comparison with others in the same age group (Fig. 3.1).
- Some older tests have used the concept of *mental age,* where tasks are graded for the age group that can be expected to pass them.
  - Although some tests still incorporate the concept, it is no longer used in direct calculation of IQ scores.
- **Wechsler intelligence tests**
  - Psychologist *David Wechsler* developed a series of three tests evaluating intellect in age groups from 4 to 75 years (Table 3.3).
  - Each test is made up of subtests in two broad categories, *verbal* and *performance:*
    - Verbal tests include measures of general information, vocabulary, and verbal reasoning.
    - Performance tests emphasize visual and visuospatial abilities, speed, and nonverbal problem solving (Table 3.4).
- **Other approaches**
  - The *Stanford-Binet* test is a measure of intellect for ages 2 through adult, now most commonly used in persons of limited ability.
  - Some briefer tests exist for quick estimation of intellectual ability; they sample a more limited range of abilities and vary in how well they predict scores on more comprehensive tests:
    - Kaufman Brief Intelligence Test (KBIT)
    - Slosson Intelligence Test
    - Shipley Institute of Living Scale

Table 3.4 WAIS-R Subtests

Verbal	Performance
Information	Picture completion
Digit span	Picture arrangement
Vocabulary	Block design
Arithmetic	Object assembly
Comprehension	Digit symbol
Similarities	

- Tests have also been developed for people with language limitations:
  - Test of Nonverbal Intelligence (TONI)
  - Peabody Picture Vocabulary Test—Revised

Aptitude and Achievement Tests

- *Aptitude tests* are intended to measure potential skills.
- *Achievement tests* measure what an individual has already accomplished in mastering skills:
  - The *Wide Range Achievement Test—Third edition* (WRAT-3) provides relatively brief screening of basic reading, spelling, and arithmetic skills.
  - Other measures such as the *Woodcock-Johnson* tests of achievement provide more detailed and comprehensive assessment.

Assessment of Personality and Psychopathology

- *Multidimensional inventories* evaluate a variety of personality characteristics or areas of psychiatric symptoms, usually based on how the person decides that a variety of statements apply to him or her.

  - **Minnesota Multiphasic Personality Inventory (MMPI)**
    - The MMPI is the most widely used objective personality inventory.
    - Interpretation is primarily based on 10 clinical scales evaluating depression, anxiety, somatization, indicators of psychosis, social functioning, and so on (Table 3.5).
    - An important feature of the MMPI is the *validity* scales, which can identify when the subject is exaggerating, overly defensive, or erratic in responding.
  - **Million Clinical Multiaxial Inventory (MCMI)**
    - The MCMI is not so commonly used as the first is but is shorter and somewhat more direct. It also has features that evaluate validity.
    - The MCMI also includes scales measuring features relevant to each of the major personality disorders.
- Other inventories, such as the *California Personality Inventory* (CPI), are designed to evaluate personality characteristics and tendencies without the clinical and diagnostic emphasis of tests such as the MMPI and MCMI.

Table 3.5 MMPI Clinical Scales

NAME		DESCRIPTION
Hy	Hypochondriasis	Preoccupation with and exaggeration of physical symptoms
D	Depression	Disturbed mood
Hs	Hysteria	Dramatic behavioral tendencies
Pd	Psychopathic deviate	Antisocial behavior and substance abuse
MF	Masculinity—Femininity	Identification with gender stereotypes
Pa	Paranoia	Suspiciousness, sensitivity
Pt	Psychasthenia	Anxiety and brooding
Sc	Schizophrenia	Unusual thinking and experiences
Ma	Mania	Elevations in mood and activity level
Si	Social introversion	Tolerance and comfort in social relationships

■ *Specific objective measures* exist to evaluate particular characteristics or problem areas.

● Psychiatrist *Aaron Beck* developed questionnaires dealing with symptoms of anxiety and depression.

• The tests include the Beck Depression Inventory, Beck Anxiety Inventory, and Beck Hopelessness Scale.

• Although they all have the limits of self-report and the questions are obvious in content, they provide quick assessment of a range of complaints.

● Other objective tests have been developed to evaluate virtually every area of pathology.
Examples include:

• Michigan Alcohol Screening Test (MAST)
• State-Trait Anxiety Inventory (STAI)

■ **Projective Tests**

The hallmark of *projective* tests is the use of *ambiguous* stimuli. Because stimuli do not have obvious objective meaning, the subject must interpret them on the basis of his or her own motives, needs, and perceptions. The absence of objective meaning is believed to help circumvent conscious defenses.

● **Rorschach Ink Blot Test**

• This test is the most widely used projective personality test, used to evaluate the quality of thinking and major defense mechanisms.

• Stimuli consist of 10 cards, each containing a bilaterally symmetrical ink blot. Half are monochromatic, and half include varying amounts of color.

• The subject identifies what the inkblot looks like, and interpretation depends on the content and perceptual quality of associations made as well as how much and what part or parts of the blot are used.

● **Thematic Apperception Test (TAT)**

• Stimuli consist of cards depicting persons engaged in ambiguous activities, and the subject makes up a story explaining what is happening.

• Responses are believed to be reflective of the subject's motivations, needs, and interpersonal conflicts.

- Sentence Completion Test
    - Stimuli consist of sentence "stems," such as "The best time . . . " or "Friends are . . . ."
    - Responses are typically interpreted in terms of expressions of conflict, unfulfilled wishes, and needs.

NEUROPSYCHOLOGIC ASSESSMENT

Neuropsychologic assessment is the use of tests of intellectual, cognitive, social, and emotional functioning to evaluate the integrity of the brain.

- **Purposes**
    - Evaluation of impairments secondary to injury, illness, medication, and so on.
    - Diagnosis of conditions in which behavioral manifestations precede observable structural change in the brain (e.g., dementia of the Alzheimer's type), or where structural change cannot be seen (e.g., mild head trauma).
    - Neuropsychologic batteries may be of assistance in localizing brain lesions, but this function has receded in importance with developments in neuroimaging technology.

- **Functions Evaluated**
    - General psychologic functioning
        - Intellect
        - Academic skills
        - Personality and psychopathology
    - Neuropsychologic functions
        - Sensory and motor skills
        - Attention and concentration
        - Mental speed and flexibility
        - Language
        - Memory
        - Visuospatial and constructional skills
        - Reasoning and problem solving

- **Neuropsychologic Batteries**

    There are dozens of tests available for evaluation of specific functions. It is common for neuropsychologists to employ standardized *batteries,* or collections of procedures evaluating a variety of capabilities.

    - *Halstead-Reitan Battery* is probably the most commonly used battery.
        - It consists of seven tests for evaluation of aspects of visual, auditory, and tactile abilities.
        - It usually is supplemented with tests of memory, intelligence, and language.
        - Performance is evaluated in terms of:
            - Level of performance
            - Pathologic patterns of performance
            - Lateralized deficits
            - Presence of pathognomonic signs

> **Box 3.1**
>
> **CALCULATION OF ODDS AND RISKS**
>
> *Example:* Two hundred patients are randomly assigned to either a treatment group or a placebo group to study effectiveness of the treatment in preventing relapse. Results are below.
>
> | | Relapse | No relapse |
> |---|---|---|
> | **Treatment** | 25 | 75 |
> | **Placebo** | 50 | 50 |
>
> Odds of relapse in the treatment group are 25/75, or 1/3.
> Odds of relapse in the placebo group are 50/50, or 1/1 (even odds).
> Risk of relapse in the treatment group is 25/100, or 25%.
> Risk of relapse in the placebo group is 50/100, or 50%.
> Relative risk of relapse in the placebo group is 50/25, or 200%.
> Attributable risk of relapse as a result of not receiving the treatment is (50-25)/100, or 25%.

- **Luria-Nebraska Battery**
  - It consists of just over 260 items testing various aspects of sensory, perceptual, and cognitive ability.
  - Its items contribute to a variety of scales that reflect specific areas of ability and functioning in different areas of the brain.

EPIDEMIOLOGY AND STATISTICS

Epidemiologic Concepts

- *Epidemiology* is the study of frequency and pattern in the occurrence of disease.

- **Incidence and prevalence**
  - *Incidence* refers to the number of *new* cases of a disease appearing in a given period of time.
  - *Prevalence* refers to the number of *existing* cases of a disease at a particular *point* or *period* of time.

- *Odds* and *risk* are alternative ways of expressing the likelihood that events (such as disease, recurrence, complications) will occur. Events must be dichotomous in nature.
  - *Odds* are calculated as a comparison of those falling on either side of the dichotomy (e.g., those who relapse versus those who do not).
  - *Risk* is calculated as a comparison of those falling in one category (e.g., those who relapse) with the total number of subjects (Box 3.1).
    - *Relative risk* compares those who fall in a category under two different conditions (e.g., those who relapse after treatment or after no treatment).

Chapter 3 Measurement of Behavior

- *Attributable risk* is calculated as the proportion of patients who relapse without treatment *minus* those who relapsed even with treatment.

Basic Statistical Concepts

- **Dependent and independent variables**
    - A *variable* is any characteristic or condition that can have more than one value.
    - An *independent variable* is a condition directly controlled by the experimenter (e.g., amount of medication administered).
    - A *dependent variable* is a condition whose value presumably depends on the value of the independent variable (e.g., a change in symptoms as a result of the amount of medication administered).

- **Measures of central tendency**
    - The *mean* is the arithmetic average of a set of values or scores calculated as the sum of scores divided by the number of measurements.
    - The *median* is the score dividing in half a set of scores arranged from lowest to highest.
        - 50% of scores fall below the median
        - 50% of scores fall above the median
    - The *mode* is the single most frequently occurring value or score.

- *Measures of variability* reflect the extent to which observed values deviate from the central tendency.
    - To calculate *variance,* the deviation of each score from the mean is squared, and the resulting values are summed and then divided by the number of scores (minus 1).
    - The *standard deviation* is the square root of the variance.

- *Distribution* refers generally to the patterns in which values of a variable occur in a population.
    - The *normal distribution* is a theoretical model in which scores or values vary in a symmetrical fashion from the mean (Fig. 3.2).
        - The majority of scores (68%) fall within ±1 standard deviation from the mean.
        - Measures of central tendency (mean, median, and mode) have the same value in a normal distribution.
        - Also referred to as a *bell-shaped* or *gaussian* distribution.
    - Not all characteristics are normally distributed.
        - In a *negatively skewed* distribution, scores cluster at the higher end of the scale.
        - In a *positively skewed* distribution, scores cluster at the lower end of the scale.
        - In a *bimodal* distribution, scores cluster around two distinct peaks.

- **Hypothesis testing** Hypotheses are statements about expected relationships between independent and dependent variables (Box 3.2).
    - The *null hypothesis* is the assumption that no relationship exists be-

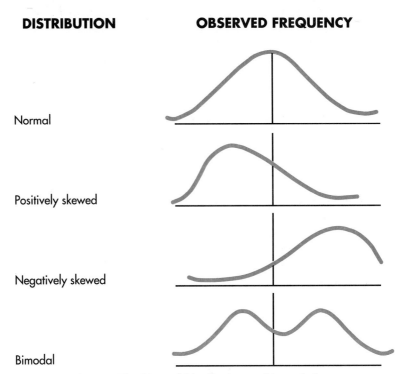

Fig. 3.2 The Normal Distribution and its variations

Box 3.2

HYPOTHESIS TESTING	
Hypothesis	Medication *x* reduces risk of relapse in Disease *y*.
Null hypothesis	Medication *x* has no effect on relapse in Disease *y*.
Type I error	Experimenter incorrectly concludes that Medication *x* reduces relapse.
Type II error	Experimenter incorrectly concludes that Medication *x* has no effect on relapse.

tween the dependent and independent variables. The essence of hypothesis testing is deciding whether the results obtained do or do not justify rejecting the null hypothesis.

- *Type I and type II errors*
  - *Type I error* occurs when the null hypothesis is incorrectly rejected.
  - *Type II error* occurs when the null hypothesis is incorrectly confirmed.
- *Statistical significance* is an expression of how likely it is that observed effects or relationships were attributable to chance.
  - Significance is expressed as a *p*, or probability value, interpreted as the chance of a type I error.

- The conventional significance level of $p < 0.05$ means a very low risk of incorrectly rejecting the null hypothesis (5 or less times in 100 cases).
- Whether a difference or effect is significant is in part dependent on the number of subjects. A higher level of significance may be required with high numbers of subjects or multiple statistical tests.

Statistical Analysis

- A *t-test* is used to compare mean scores from two groups or from the same group on two different occasions.
  Example: Compare mean weight gain in anorexic patients treated with medication A or a placebo.

- *Analysis of variance* tests differences between mean scores from more than two samples.
  - *One-way* analysis of variance tests differences between mean scores from groups divided along one variable.
  Example: Compare the mean weight gain in anorexic patients treated with medication A, medication B, or a placebo.
  - *Two-way* analysis of variance tests differences between mean scores from groups divided along two different variables.
  Example: Compare the mean weight gain for anorexics, dividing subjects not only by medication status but also by onset before and after 21 years of age.

- *Correlation* expresses the relationship between two continuous variables.
  Example: What is the relationship between weight gain and the number of milligrams of medication A?
  - Correlation coefficients vary from -1.0 to $+1.0$.
  - In a *direct* relationship, variable 1 goes up or down in the same direction as variable 2. The coefficient is positive.
  - In an *inverse* relationship, the coefficient is negative. As one variable goes up, the other goes down.

- *Multiple regression* allows one to calculate the correlation between one variable (the dependent variable) and more than one other variable (independent variables).
  Example: What is the relationship between weight gain and the combination of dosage of medication A and varying numbers of psychotherapy sessions?

- *Chi-square* allows one to test for differences in the frequency of group membership.
  Example: Does the percentage of anorexics with onset before or after 21 years of age differ in subjects from rural versus urban areas?

- *Nonparametric tests* include a variety of procedures used when the phenomena studied are not normally distributed or come in an unusual form (e.g., rankings).

Research Design

- *Validity and reliability* apply to experimental design as well as to specific tests.
  - Designs are intended to minimize threats to validity, or factors that might confound findings and limit the conclusions that can be drawn from the data.

- Experimental effects must be replicable.
- *Controls* are procedures designed to eliminate sources of bias or confusion about the meaning of experimental results.
  - *Randomization* means that subjects have an equal chance of being assigned to any of the experimental groups or conditions.
    - It is essential to a true experiment in ensuring that groups in a study differ *only* in group membership.
    - It controls for bias in group assignment (e.g., assigning sicker patients to the treatment group).
  - *Control groups* provide a basis of comparison.
    - A *no-treatment control group* is necessary in some cases to show that a treatment produces more benefit than the simple passage of time does.
    - A *placebo control group* is necessary in some cases to show that treatment produces benefit beyond what might occur just because the subject is receiving any kind of treatment or attention.
  - *Blind procedures* control for the possible bias introduced if experimenters or subjects know about group membership. In *double-blind studies,* neither subjects nor experimenter know which subjects are taking which medication or placebo.
  - *Crossover* designs reverse subject assignment to placebo and treatment groups so that each subject participates in each condition. Finding the same difference between the treatment and the placebo increases confidence that the difference is attributable to the effect of the treatment and not to unrecognized differences between the groups.
- **Prospective and retrospective designs**
  - *Retrospective* studies examine events that have already occurred.
  Example: Charts are reviewed to see if anorexics getting group therapy and medication did better than those getting medication alone.
  - *Prospective* studies are set up before the events occur.
  Example: Newly admitted anorexic patients are randomly assigned to either medication alone or medication plus group therapy.
- **Cross-sectional and longitudinal designs**
  - *Cross-sectional* designs study phenomena in people differing in age or stage of illness.
  Example: A change in intellect with age is studied by comparison of test scores of 20-year-old, 40-year-old, and 60-year-old subjects.
  - *Longitudinal* studies follow a single cohort over time.
  Example: A change in intellect with age is studied by testing of a group of 20-year-old subjects and then testing of the same group again at 40 and 60.
- *Meta-analysis* is a procedure for examining the general conclusions to be derived from a body of studies rather than a single study.

Multiple Choice Review Questions

1. Test results come back indicating that your patient does not have the illness in question. Your confidence in this conclusion depends on the _____ of the test.
   a. Face validity
   b. Positive predictive value
   c. Sensitivity
   d. Specificity

2. The important characteristic of *projective* personality tests is which of the following?
   a. Inclusion of scales evaluating the validity of responses
   b. The concept of mental age
   c. The use of ambiguous stimuli
   d. Evaluating potential for developing specific skills

3. A patient is evaluated with the Halstead-Reitan Battery. The most likely reason for this evaluation is which of the following?
   a. Determining the effects of a cerebrovascular accident
   b. Establishing a diagnosis of mental retardation
   c. Differentiating between two possible personality disorders
   d. Determining the patient's ability to complete a college degree

4. A researcher wishes to split her sample in half so that half the subjects fall above a certain score and half fall below. The researcher needs to identify which of the following?
   a. Mean
   b. Median
   c. Mode
   d. Variance

5. You are studying the effectiveness of three different dosages of a medication to reduce blood pressure. The appropriate statistical test is which of the following?
   a. Chi-square
   b. T-test
   c. Analysis of variance
   d. Multiple regression

Part 2

Development and the Life Cycle

Chapter 4

Pregnancy and Infancy

PREGNANCY

Emotional Adaptations to Pregnancy

- Pregnancy is a developmental challenge in which the expectant couple must adapt to new roles and changes in their relationship.

  - Most pregnancies are planned.
  - Ambivalence and conflict may arise when new roles are seen as threatening or overwhelming.
  - There must be sufficient self-esteem to see oneself as capable of being a good parent.
  - Emotional influence of pregnancy is colored by all the elements of context, including personal experience, personality, beliefs, life circumstances, and maturity.

- Maternal adaptation

  - Physical changes may be an affirmation of womanhood and adulthood, or a threat to perceived attractiveness.
  - The expectant mother must develop an *affiliative response to the fetus* in which she comes to see her interests and those of the fetus as coinciding.
  - Changes in emotionality, including *lability,* may be observed.

- Parents have fantasies and fears about their infant, and these must be reconciled to the reality of the infant actually born.

The Developing Pregnancy

- The first trimester is characterized by *morning sickness,* and the pregnancy is often more real to the woman because of the physical factors.
- *Quickening* is an important feature of the second trimester. The detectable movements of the fetus contribute to the parents' coming to see the fetus as a separate individual.
- The third trimester
  - Practical preparation and decision making
  - Physical stress and pressure for the woman
  - Feelings of eagerness and anxiety

Sexuality

- Some women feel increased sexual desire, whereas others feel less desire, reflecting physical changes and perhaps fear of damaging the fetus.
- Fathers may be hesitant sexually because of the mother's physical and emotional changes, or may also fear harming the fetus.
- Changes in sexual activity can be a strain on the marriage. Fathers

Box 4.1

BENEFITS OF PREPARATION FOR CHILDBIRTH

Shorter labor
Reduced need for analgesia
Fewer intrapartum complications
More positive perception of the infant
Greater likelihood of sustained breast feeding

who have extramarital affairs are most likely to do so in the third trimester.

- **Other Issues**
  - Teenage pregnancy
    - Almost half of females 15 to 19 years of age are sexually active, and about 30% of these will become pregnant.
    - About 25% of these pregnancies are voluntarily terminated.
    - Pregnant teenagers are less likely to receive proper prenatal care and are at greater risk for complications in childbirth.
  - Prenatal care
    - It is essential to manage maternal conditions, such as hypertension and gestational diabetes.
    - Lack of prenatal care is associated with higher infant morbidity and mortality.
    - Proper prenatal care is less likely with lower socioeconomic status, lack of education, and adolescent mothers.

  - Substance abuse
    - Maternal smoking is associated with higher risks for spontaneous abortion, low birth weight, and respiratory infections in the child.
    - Maternal alcohol use puts the child at risk for *fetal alcohol syndrome.*
      - The full syndrome is characterized by mental retardation, facial abnormalities, growth deficiencies, and persistent social and behavior problems.
      - Children who do not display the classic syndrome may have less dramatic problems variously described as *fetal alcohol effect* and *alcohol-related behavior disorders.*
    - Maternal opiate use can result in neonates displaying withdrawal syndromes.
    - Maternal use of cocaine, particularly *crack,* is associated with neonatal withdrawal and probably also with persisting cognitive and behavioral problems.

CHILDBIRTH

Preparation for childbirth (e.g., *Lamaze* classes) has many beneficial effects, particularly when preparation emphasizes maternal control (Box 4.1).

- Delivery by *cesarean section* has increased greatly in the past two decades.
  - Birthing is perceived as less positive and more stressful.

Box 4.2

> **FACTORS ASSOCIATED WITH BETTER PARENTAL COPING WITH PREMATURE BIRTH**
>
> Actively seeking information
> Expressing negative feelings
> Stable marriage
> Adequate income and insurance
> Supportive social and interpersonal network
> Supportive NICU staff
> Residing near the NICU

NICU, Neonatal intensive care unit.

- Disappointment and depression are more common in mothers who planned natural births.
- Infants and their capabilities are evaluated more guardedly by parents though the infants do not differ objectively from vaginally delivered infants.
- Fathers may be more involved in early caregiving.

■ *Premature birth* can provoke reactions of sorrow, guilt, and fear in parents.

- Aspects of parental response and resources are found to be predictive of better coping with the stresses involved (Box 4.2).
- One consequence may be development of a *vulnerable child syndrome* in which the child is perceived to be more vulnerable to accidents and illness and is overprotected as a result.

■ *Stillbirths and neonatal deaths* produce grief reactions as fully intense as those provoked by other interpersonal losses.

- Grieving mothers frequently report benefit from opportunities to see and hold their infants or have a picture.
- Parents often derive comfort from funeral services and other death practices (e.g., burial with a marker) that acknowledge the loss of a person.

■ *Postpartum reactions* can range from brief and transient dysphoria (the relatively common *postpartum blues*) to the relatively rare *postpartum psychosis*.

- As many as half of new mothers may be briefly depressed, reflecting stress and fatigue, hormonal changes, and anxiety about parenting.
- Mood disturbance may reach the level of *major depression* in as many as 1 in 10 new mothers.
- A small percentage of those who become seriously depressed will exhibit symptoms of psychosis, such as hallucinations or delusions.
- Psychotic symptoms are associated with increased risk of suicide or infanticide, or both.
- The risk of postpartum mood disorders is higher when new mothers are isolated, lack experience, or have previous history of postpartum reactions.

Neonatal and Infant Behavior

■ **Infant Competence**

- Physical capabilities and reflex behavior
  - Infants are born with a rich repertoire of reflex behaviors that appear

during gestation and disappear in the healthy infant during the first year of life (Table 4.1).

• Persistence of reflex behaviors beyond the time when they normally disappear can indicate neurologic impairment.
Example: Persistence of the tonic reflex past 6 months of age is an indicator of cerebral palsy.

• Infants are typically able to sit unassisted at 6 months and walk at 12 months.

- **Social responses**
  - Newborns are participants in the developing relationship with the mother.
    - They respond differentially to the sound of the mother's voice.
    - They respond differentially to speech.
  - Feeding and sleep are disrupted when the infant is deprived of maternal facial expression and verbalizing.
  - Being lifted to the caregiver's shoulder quiets the crying infant and increases alertness.

- **Perceptual skills**
  - Tactile sensation is developed earliest and most fully at birth.
  - Newborns respond differently to sweet and noxious odors, and at 6 days post partum they can reliably recognize the scent of their mothers' breast pads.
  - Newborns can visually discriminate and respond differentially to human faces.
  - Neonates demonstrate physiologic, biochemical, and behavioral responses to pain.

- **Neonatal Assessment**
  - *Apgar* scores are recorded at birth and allow evaluation of physiologic parameters, including heart rate, muscle tone, and respiratory effort.
  - *Dubowitz* scores are based on the evaluation of reflex behaviors and can help to establish gestational age.
  - *Brazelton* scores are based on the assessment of reflexes and responsiveness to stimuli.

Table 4.1 **Infant Reflexes**

Reflex	Age of Disappearance (months)
Rooting	3-4
Moro	3-4
Stepping	4-5
Tonic neck	4-6
Grasp	5-6
Babinski	12-14

Chapter 4 Pregnancy and Infancy

- **Attachment and Reciprocity**
  - Attachment to the primary caregiver or mother is a development of a primary and intimate relationship.
    - The process is reciprocal, in that both infant and parent are active participants.
    - Infants become distressed when mothers are unresponsive in movement, facial expression, and vocalization. These findings have implications for how infants are affected by maternal depression.
  - Disturbances in attachment may have dramatic implications for development.
    - Sustained separation from the primary caregiver in infancy is associated with risk of *anaclitic depression* characterized by withdrawal and failure to thrive.
    - *Harry Harlow's* studies of monkeys show that rearing in isolation disrupts normal social behavior.
      - Mating and maternal behavior are adversely affected.
      - Rehabilitation is possible if isolation does not persist past 6 months.

- **Crying and Colic**
  - Crying typically peaks at about 6 weeks.
    - Crying is differentiated according to the stimulus causing it (such as hunger versus wet diaper).
    - There is an objective basis to parent reports that they can distinguish cries by cause.
  - *Colic* is unexplained and excessive crying between 3 weeks and 3 months of age.
    - Infant behavior is suggestive of gastrointestinal (GI) distress, but no specific GI abnormalities have been reliably identified.
    - Colic is believed to reflect difficulty in self-regulation of arousal.

- *Temperament* refers to behavioral features and characteristics along which infants differ (Box 4.3).
  - Temperamental characteristics influence the match between infant and caregiver.

Box 4.3

CHARACTERISTICS OF TEMPERAMENT

Mood
Level of activity
Intensity of responses
Threshold of responsiveness
Adaptability to change
Distractibility
Attention span and persistence
Rhythmicity in biologic and behavioral patterns
Approach or withdrawal in response to new stimuli

- Temperamental characteristics predict some later behavioral tendencies.

Problem Infants

- *Prematurity* is associated with risk of central nervous system insult and later developmental problems.
    - Learning disorders, hyperactivity, and emotional disturbances are more common.
    - There is increased risk of child abuse.
- *Birth defects* can evoke depression, shame, and other emotional responses from parents who adversely affect attachment.

Multiple Choice Review Questions

1. During pregnancy which of the following is *true?*
   a. Quickening is an important feature of the first trimester.
   b. Morning sickness is a major stressor in the final trimester.
   c. Women may feel more or less sexual desire.
   d. Emotional changes and lability in women indicate severe stress.

2. Which of the following is *true* after the birth of a child?
   a. About half of new mothers will be diagnosed with major depression.
   b. Even brief depression is a psychiatric emergency.
   c. New mothers are unusually resistant to disturbance of mood.
   d. Isolated and inexperienced mothers are more likely to be depressed.

3. Which of the following statements about infant crying is *correct?*
   a. Parents accurately report that crying differs depending on the reason for crying.
   b. Colic is the term for normal infant crying that peaks at about 6 weeks.
   c. Colic is crying due to gastrointestinal distress.
   d. Normal crying peaks at about three months of age.

4. In evaluating the perceptual capabilities of newborns, which of the following is *true?*
   a. No distinctive or organized pain response can be identified.
   b. Newborns cannot distinguish between human faces and random patterns.
   c. Tactile sensation is the most fully developed at birth.
   d. Newborns cannot make even rudimentary olfactory discriminations.

5. All of the following statements about mother-infant interaction are correct *except* which of the following?
   a. Very early after birth, infants respond specifically to the mother's voice.
   b. Lack of caregiver attention is associated with anaclitic depression.
   c. Infants are more rapidly calmed if the mother keeps silent and avoids eye contact.
   d. Babies become quiet and alert when lifted to the caregiver's shoulder.

Chapter 5

Childhood and Adolescence

EARLY CHILDHOOD (1 TO 6 YEARS)

- **Physical Development**
  - Growth
    - The average child triples in weight and grows 24 to 30 cm in the first year and reaches half of adult height between 2 and 3 years of age.
    - During the preschool years the average child gains about 2 kg and grows 6 to 8 cm per year.
  - Motor skills
    - *Fine motor skills* involve control over specific muscles, particularly in the hands and fingers, and coordination of hand and eye (e.g., coloring with a crayon).
    - *Locomotion* skills are the major development in *gross motor skills.*
    - There is considerable variation in the ages at which normally developing children develop various motor skills. Table 5.1 lists the typical ages at which various abilities are established.
  - *Toilet training* typically becomes a parental concern at about 2 years of age.
    - Voluntary sphincter control typically develops between 18 and 24 months.
    - Bladder control usually develops later (30 to 36 months), first for the daytime and later for the night.
    - Fifteen to twenty percent of 5-year-old children continue to have *nocturnal enuresis* (bedwetting); by 10 years of age the rate is only 5%.
    - Toilet training is more easily undertaken if the child has some language skills.

- **Cognitive Development**
  - *Piaget's* theory describes a series of stages driven by two complementary processes.
    - *Assimilation* occurs when new information and experiences are incorporated into existing patterns of thinking and behavior.
    - *Accommodation* occurs when experiences require modification of existing patterns of thinking and behavior.
  - In the *sensorimotor stage* (up to 2 years) the child learns about the environment through motor activity and sensory experience.
    - Between 12 and 24 months there is increasing understanding of *object permanence;* there is recognition that objects exist even when not immediately visible.

Table 5.1 *Motor Skill Development*	
MOTOR SKILL	TYPICAL AGE OF ACQUISITION
Standing	9-10 months
Walking without assistance	12-15 months
Walking up stairs with both feet on each step	2 years
Balance for several seconds on one foot	3 years
Walking downstairs with one foot on each step	4 years
Skipping on both feet	5 years

- Object permanence shows the developing ability to use mental representations of objects in thinking.
- In the *preoperational stage* (2 to 6 years) the child begins to think symbolically and with mental representations of events.
  - Thinking is *egocentric* in that the child cannot see things from another's perspective.
  - *Causality* is assumed when events occur at the same time (e.g., a boy thinks he got sick because of naughty behavior).
  - The child is still dependent on perceptual evaluation of the environment (e.g., a container that looks bigger is assumed to hold more even if the amounts are the same).

Language

- Vocabulary expands from the first few words at 1 year of age to approximately 300 words at 2 years and about 1000 words at 3 years.
- Articulation, sentence length, and sentence structure gradually approximate adult levels.
  - Earliest sentences take a form called *telegraphic speech* consisting of two or three words without adjectives or prepositions (e.g., "Me want drink").
  - Rules, such as adding "-ed" to indicate past tense, are used to excess ("We goed outside").
  - About 4 years of age children can ask questions in adult form and begin to use more complex grammatical forms, such as future tense.
- *Receptive* language develops more rapidly than *expressive* language; children understand more than they can express.

Social and Emotional Development

- Play
  - Play from 12 to 24 months includes use of objects in imitation of their function.
  - Before 3 years of age children together engage in *parallel play* in which they may share materials but work independently.
  - From about 3 years play becomes increasingly cooperative and incorporates fantasy elements about adult roles.
- *Separation anxiety* peaks at about 18 months, though regression is not remarkable when children are stressed or anxious.

- Fears and fantasies
  - *Nightmares* and *transient phobias* (fear of monsters or the dark) are common in preschool children.
  - About half of preschool children will have *imaginary companions* for a time.
- **Sexuality Issues**
  - *Sex identity* is consolidated between 2 and 3 years of age.
  - Preschool children are interested in physical differences and may pursue their interest in play (e.g., playing doctor) and in questions asked of parents.
  - Preschool children commonly explore their genitalia, and interest in masturbation is not unusual.

School-Aged Children (6 to 12 Years)

- *Physical Development* is characterized by steady growth and improvement in both gross and fine motor skills. The skills of riding a bicycle, skipping rope, and playing ball games are commonly acquired.
- **Cognitive Development**
  - Characterized by the stage of *concrete operations* (Piaget).
    - Capacity for rule-governed manipulation of information, such as calculations or serial ordering.
    - *Classification* skills are developed and games have increasingly complex rules.
    - The principle of *conservation* is acquired; e.g., children recognize that a quantity is not changed by containers of different shapes.
    - Systematic problem solving and deductive reasoning are observed.
  - Atypical development often first becomes evident with the structured demands of school.
    - *Mental retardation,* characterized by *subnormal intellect* and *adaptive deficits*
    - *Learning disabilities,* where intellect is in the normal range but academic functioning is below expectations

- **Social Development**
  - Segregation of play groups by sex is typical.
  - Relationships expand to include those outside the family.
  - Challenges to self-esteem as children are faced with succeeding in school and fitting in socially.

Adolescence (12 to 18 Years)

- *Puberty* is the defining biologic feature.
  - *Secondary sex characteristics* develop.
  - Onset of *menarche* in girls is around 12 years of age and *first ejaculation* in boys is about 2 years later.
  - Growth spurts are also observed earlier in girls.

Chapter 5 Childhood and Adolescence

- Maturation begins earlier in girls, but by late adolescence boys are typically larger and stronger.
- Relatively early maturation has more benefits in self-esteem and confidence for boys.

■ *Cognitive Development* features the emergence of *formal operational thinking*.

- Capacity for abstract thinking and thinking in hypothetical terms develops.
- Ability to think about ideas, often idealistic and absolute, develops.

■ **Social and Emotional Development**

- There is a greater influence of peers and a lesser family influence.
- Dating, crushes, and sexual urges begin.
- Personal identity forms.
- Preparation begins for adult roles in the spheres of vocation and marriage.

■ **Problems of Adolescence**

● Sexuality and teenage pregnancy

- Almost half of female adolescents 15 to 19 years of age are sexually active and about 30% become pregnant.
- Early sexual activity is associated with more limited intellect, lower academic achievement, and lack of educational goals.
- Contraceptive use is erratic.
- Risks of pregnancy and sexually transmitted diseases must be addressed in education and counseling of adolescents.

● Substance abuse

- Alcohol use is the most common problem and it contributes strongly to accidental deaths, principally in motor-vehicle accidents.
- Involvement with drugs and alcohol may be evident in apathy, poor school performance, and lack of participation in activities appropriate for age.

● Conflict with parents and difficulty with communication

- Adolescents may have strong feelings of alienation and isolation.
- The drives toward independence may alarm parents and provoke conflict.

● The leading cause of death among adolescents is *accidents*.

Multiple Choice Review Questions

1. Telegraphic speech is which of the following?
   a. An early sign of mental retardation
   b. Characteristic of the preoperational stage of cognitive development
   c. Used by children during play
   d. Simplified speech typical of early language development

2. In your examination of a 28-month-old child, you are concerned about which of the following?
   a. The child plays alongside other children but the play activities are not yet really collaborative.
   b. The child has not yet developed voluntary bowel control.
   c. The child has developed daytime but not nocturnal bladder control.
   d. The child climbs stairs one at a time with both feet placed on each step.

3. Which of the following is *true* in the preoperational stage of cognitive development?
   a. A child is unable to see things from the perspective of another person.
   b. A child has mastered the concept of conservation.
   c. A child has not yet acquired the concept of object permanence.
   d. A child learns through physical manipulation and sensory experience.

4. By the time children reach school age which of the following is *true?*
   a. Cognitive development features the emergence of formal operational thinking.
   b. Boys are bigger and physically superior to girls.
   c. Systematic problem solving is still not part of a child's repertoire.
   d. They tend to play in groups that are segregated by sex.

5. Among adolescents which of the following is *true?*
   a. The leading cause of death is suicide.
   b. The most common substance abuse problem involves alcohol.
   c. The more intelligent and academically successful are more sexually active.
   d. Cognitive development is dominated by the concrete operations stage.

Chapter 6

Early and Middle Adulthood

DEVELOPMENTAL FEATURES

Adulthood is somewhat arbitrarily divided into *early (18 to 40)* and *middle (40 to 65)* periods.

- Early adulthood
  - Peak levels of development in physical and mental characteristics occur.
  - Key challenges focus on *development of intimate relationships* and *work* or *career.*
- Middle adulthood
  - There is probable involvement of peak levels of affluence, power, and accomplishment.
  - *Key challenges*
    - Dealing with signs of physical aging
    - Maintaining a sense of productivity (Erikson's *generativity*)
    - Relationships with adult children and aging parents
    - *Diversity* is a prominent feature of adult life; people vary in the timing of events, such as marriage and childbearing, and may revisit some tasks in remarriage and career changes.

FAMILY AND RELATIONSHIPS

Marriage

- Only 5% of adults in the United States will never marry.
- Good marriages are associated with multiple benefits.
  - Lesser likelihood of depression
  - Higher self-esteem
  - Better physical health
- The benefits of marriage are so substantial that they overcome the fact that married persons are at risk for a higher number of undesirable life experiences.
- *Cohabitation* before marriage does not increase the chances of success. Marital adjustment is poorer, divorce is more likely, and physical abuse is more common among couples who cohabitate.

Divorce

- Nearly half of marriages in the United States end in divorce.

Table 6.1	Predictors and Effects of Divorce
Predictors of divorce	Family history of divorce Infidelity Substance abuse Premarital pregnancy Teenage marriage
Effects on children (increased risk)	Suicide and depression Divorce Substance abuse Poor school performance

- Divorce is associated with higher risks for both physical and psychiatric illness.
- The majority of divorces involve children under 18 years of age, and there are multiple adverse effects on the children (Table 6.1).
- Divorce risk is highest in the first 5 years, but peaks again in the fifteenth and twenty-fifth years.
  - Divorce in midlife is frequently linked to infidelity.
  - 85% of divorced men remarry, but chances of remarriage among women diminish with increased age and education.

Parenting

- Almost *90% of married couples have children.* About half of couples without children are childless by choice.
- The arrival of children is commonly a source of great joy and satisfaction but also a source of anxiety and strain.
  - Couples may have fears about their adequacy as parents.
  - Parents must adapt to changes in their social lives and opportunities for intimacy as a couple.
  - Couples often come to divide tasks along the lines of more conventional sex roles.
- Parenting tasks change as the child grows.
  - Parenting involves "letting go" as children move from the dependency of infancy to greater involvement with the world at large.
  - *Sexuality* is typically the primary concern for parents of adolescents.
- Single-parent families
  - About *85% are headed by females.*
  - Strains are associated with more *limited income* and lack of *support from a partner.*
  - Compared with married women and single women living alone, *unmarried mothers are more often ill and utilize health care more frequently.*

WORK AND LEISURE

Importance of Work

- Work fulfills multiple needs apart from providing income (Box 6.1).

Chapter 6 Early and Middle Adulthood

Box 6.1

MULTIPLE ROLES OF WORK

Providing income
Structuring time
Defining social status
Defining personal identity
Providing social interaction
Providing opportunities for expression

Box 6.2

BENEFITS OF EXERCISE

Reduced depression
Reduced anxiety
Increased stress tolerance
Increased self-esteem
Improved sleep
Improved concentration

- Work satisfaction is correlated with longevity.

Work and Health

- About 400,000 new cases of work-related illness occur each year.
  - 90% of family practitioners encounter work-related illness or injury at least once per week.
  - Only 25% of physician reports mention occupation or employment status.
- *Job stress* has been correlated positively with *risk of heart disease.*
- *Job loss* is associated with increased risk of *physical illness, substance abuse, depression, and disrupted family life.*

Work and Sex

- In the workplace, *segregation by sex* has declined less than segregation by race in the last three decades.
- Women who work in traditionally "female" occupations report more depression, lower self-esteem, and lower job satisfaction than those who don't.
- Women employed outside the home still carry the bulk of responsibility for homemaking tasks in most homes.

Exercise

- Regular exercise assists in controlling blood pressure and reduces risk of heart disease.
- Regular exercise has multiple psychologic benefits (Box 6.2).

MIDLIFE ISSUES

Menopause and the Climacterium of Both Sexes

- *Menopause* is defined by the cessation of menstruation usually occurring gradually somewhere between 45 and 55 years of age.

Box 6.3

HEALTH BENEFITS OF RELIGIOUS COMMITMENT

Lower risk of depression
Lower blood pressure
Lower mortality among the elderly
Lower incidence of suicide
Lower risk of substance abuse
Lower divorce rates and higher marital satisfaction
Better response to psychiatric treatment

- Some women experience psychologic distress (e.g., depression), but most will not encounter significant problems.
- Sexuality may be more relaxed with concerns about conception eliminated.
- The male climacterium is more nonspecific physiologically and more psychologic.

■ Some parents experience an *empty-nest syndrome* of sadness and sense of loss when the youngest child leaves home.

- The syndrome is most likely among women who have limited involvement outside the home.
- Many parents enjoy the freedom they gain.

■ *Midlife crisis* typically involves confronting realities of mortality and unmet goals.

- Significant changes in work or family relationships are common precipitants.
- Midlife crisis is more likely among persons with previous adjustment problems.

■ **Midlife and Health**

- Nearly 80% of middle-aged men and women describe their health as good or excellent. Chronic illnesses, however, often first appear in middle age and require adaptation.

Religious Commitment

Religious commitment has positive influences on both physical and mental health (Box 6.3).

Multiple Choice Review Questions

1. Compared to couples who do not cohabitate, couples who live together before marriage are which of the following?
   a. Less likely to experience domestic violence
   b. Less satisfied with their sexual relationship
   c. More likely to describe the marital relationship in positive terms
   d. More likely to divorce

2. Which event of adult life best illustrates the developmental achievement of generativity?
   a. Marriage
   b. Having children
   c. Menopause
   d. Equal division of household tasks

3. All of the following statements about divorce are correct *except* which of the following?
   a. Sexual infidelity is a common cause for divorce among middle-aged couples.
   b. Only a small portion of divorced men will remarry.
   c. Risk of divorce lowers after couples have been married for 5 years.
   d. Most divorces involve children under the age of 18.

4. Which of the following statements about work and health is *correct?*
   a. The incidence of work-related illness is about 100,000 cases per year.
   b. Only 60% of physician reports specify the patient's occupation.
   c. Half of all family practitioners will encounter a work-related illness or injury in a given week.
   d. Persons who lose their jobs are at a higher risk for both physical and psychiatric illness.

5. In middle age which of the following is *true?*
   a. An "empty nest syndrome" is most common among women who have worked outside the home.
   b. Incidence of chronic illness decreases and incidence of acute illness increases.
   c. A substantial majority of persons describe their health as good or excellent.
   d. It is now known that males experience a definitive sexual change called the climacterium.

Chapter 7

Aging, Death, and Dying

CHARACTERISTICS OF THE ELDER POPULATION

- Adults over 65 years of age are an increasingly large portion of the population.
  - Americans 65 years of age or older comprised 4% of the population in 1900 and 13% in 1990 and are projected to exceed 20% by 2020.
  - At birth, life expectancy for males is 73 years and 80 years for females.
  - Males who reach 65 years of age will live on average to the age of 80, whereas females will live on average to the age of 84.
- The *socioeconomic status* of the elderly has largely improved over the course of the century.
  - The poverty rate among the elderly is about the same as in the general population.
  - One factor mitigating poverty among the elderly is the relationship between longevity and socioeconomic status; those surviving into old age are on average better off.
- Most older adults are part of family and community.
  - 95% live in the community.
  - 80% have children, and most have some contact.

CHANGES IN LATER LIFE

Changes in later life are inevitable but vary in magnitude, how quickly they occur, and how much influence they have on functioning.

- **Physical Changes (Box 7.1)**
  - Changes in strength, speed, and reaction time have noticeable effects on physical activities.
  - Changes in sensory acuity (hearing and sight) affect a broad range of activities, including reading, driving, and communication with others.
  - Changes in sleep architecture, including reduced time in deep sleep, can lead many older adults to present with complaints of insomnia.

- **Changes in the Brain**
  - *Atrophy* is observed in reduced brain weight, ventricular dilation, and sulcal widening. Although associated with dementia, the relationship is not specific enough to be of diagnostic use.
  - *Neurochemical change* includes reduced neurotransmitter concentrations, elevation of *monoamine oxidase levels,* and reduction in the activity of *choline acetyltransferase.*

Chapter 7   Aging, Death, and Dying

Box 7.1

PHYSICAL CHANGES OF AGING
Loss of brain weight
Reduced muscle mass
Osteoporosis
Loss of elasticity in connective tissue
Decline in sensory acuity
Decreased reaction time
Reduced renal functioning
Diminished cardiac output and reserves
Diminished respiratory capacity
Diminished immune response

- These changes render the elderly highly sensitive to medications with central nervous system effects.
- The elderly are highly sensitive to medications with *anticholinergic* effects and are vulnerable to development of *anticholinergic delirium.*

• Other changes include *reduction in cerebral blood flow* and *slowing on the electroencephalogram* (EEG).

Cognitive and Intellectual Changes

• Although *dementia* is clearly associated with aging, it is not the norm; even at 85 years of age, only about 25% have significant symptoms.

• Normal intellectual change is most profound in areas emphasizing speed, short-term memory, and new learning.

- Aspects of intellect that represent accumulation of experience *(crystallized intelligence),* such as fund of information, may improve well into later life.
- Early studies of intellectual change, suggestive of more generalized and significant decline, were flawed by emphasis on *cross-sectional* design and confounded age changes with cohort differences in education.
- The "use it or lose it" principle appears to apply, in that abilities are better preserved in older adults who have been intellectually active.

Sexuality

• Older adults who were sexually active when younger are most likely to be active in later life. Predictors of sexual activity in later life include *prior level of activity, health status,* and *availability of the partner.*

• Normal changes in capabilities may alter the sexual experience but do not preclude satisfying and rewarding activity (Box 7.2).

■ *Retirement* is not a distressing or traumatic experience for the average person.

• About 25% of workers are forced into retirement by illness or disability. Of these workers, half show improvement in health status after retirement.

• Among other workers, there is no increase in morbidity and mortality associated with retirement.

• Many workers will have an initial "honeymoon" period after retiring, followed by a period of disenchantment before becoming adapted to the changes in their lives.

Box 7.2

CHANGES IN SEXUAL FUNCTIONING

Males
Take longer to achieve erection
More likely require direct stimulation to achieve erection
Take longer to achieve orgasm
Longer refractory period

Females
Atrophy of the vaginal mucosa
Slower onset of lubrication
Reduced vaginal elasticity

PROBLEMS OF LATER LIFE

Health Problems

- Physical health
  - Those over 65 years of age make up about 13% of the population *but use about 33% of health care resources.*
  - About 85% *have a chronic illness,* and about 50% have some limits on their activities as a result.
  - Despite these problems, about 60% *of noninstitutionalized elderly rate their health as good or excellent,* and another 20% rate their health as fair.
  - With multiple medical problems and physiologic sensitivity to medications, *polypharmacy* becomes a significant problem area for many elderly.

- Mental health
  - In general the rate of psychiatric disorders is lower among the elderly, apart from cognitive disorders.
  - Interpersonal losses and health problems create a vulnerability to depression, and *elderly white males have the highest suicide rate.*
  - Besides depression, factors contributing to suicide include social isolation, poor health, and substance abuse.
  - 5% to 10% of persons over 65 years of age will have symptoms of dementia.
    - About half of dementia cases are attributable to *Alzheimer's disease.*
    - Careful evaluation is critical because about 15% of dementias are *reversible.*
    - Depression in the elderly sometimes mimics dementia *(pseudodementia).*

■ Growing old means coping with *losses,* including the deaths of family and friends, losses of physical health and capacities, and changes in social roles.
  - Since women live longer and tend to marry older men, *the average woman spends the last 10 years of life alone.*
  - Loss of spouse is associated with increased morbidity and mortality.

Institutionalization

- Only 5% of those over the age of 65 live in nursing homes.

Table 7.1	*Stages of Dying*
Denial	Inability to accept the diagnosis or its terminal prognosis
Anger	May be free floating; may be directed at the physician
Bargaining	Pleading for reprieve, sometimes to complete a goal or see a family event
Depression	Sadness and distress as the reality of the situation is confronted
Acceptance	Achieving a measure of serenity

- The nursing home environment can be characterized by a variety of problems:
  - Loss of privacy and control
  - Limited staff and activity resources
  - Age segregation
  - Underdiagnosis and undertreatment of psychiatric disorders

Abuse of Elders

- About 1% to 3% of elders are estimated to be victims of abuse and neglect, including unnecessary restraints and overuse of psychoactive medications.
- Perpetrators are most often spouses followed in frequency by adult children.
- Spouses may be impaired themselves.
- Many states have mandatory reporting laws similar to those dealing with child abuse.

DEATH AND DYING

Contemporary Facts of Death

- Medical advances produce compression of mortality into the later years of life.
- The definition of death has come to emphasize absence of brain functioning.
- Causes of death have shifted over the course of the century from infection to cardiovascular diseases (35%), cancer (26%), and accidents (6%).
- Infectious disease primarily kills those with compromised immune systems, such as the elderly and the chronically ill.

The Dying Process

- *Elisabeth Kübler-Ross* is well known for describing the process as a series of stages progressing from initial denial to final acceptance (Table 7.1)
  - The stages are not interpreted as a strict or progressive sequence ending definitively in acceptance for all patients.
  - Depression is common, but all terminally ill patients do not become clinically depressed.

> **Box 7.3**
>
> **RIGHTS OF THE DYING PATIENT**
>
> Right to information about the illness
> Right to informed consent regarding treatment
> Right to refuse treatment
> Right to specify in advance the procedures the individual would accept or refuse
> Right to delegate these rights to a legal surrogate

- Patients depressed enough to exhibit prolonged sadness, suicidal thinking, guilt feelings, and withdrawal can be appropriately treated with therapy and antidepressant medication.
- Patients's fears focus on *pain, loss of dignity and control, being left alone,* and the *anguish of survivors.*
- Although most people died at home at the turn of the century, 80% of deaths in 1990 took place in an institutional setting such as the hospital.
    - *Hospice care* emphasizes a multidisciplinary, family-oriented approach to assisting the dying person.
    - Hospice care may form part of care for patients in the institutional setting or may allow patients to die at home with their families.
- **Patient rights**
    - Patient rights have been articulated in law at both the federal level and in the laws of many states. Key elements are summarized in Box. 7.3.
    - The core of patient rights is promoting optimal participation in making decisions.
    - The principle of *proportionality* emphasizes that appropriate treatment is based on balancing the costs and benefits *as perceived by the patient.*

Bereavement and Grief

- *Grief* is the emotional and physical experiences of bereavement, part of the more complex process of mourning.
- Characteristics of grief:
    - Emotional distress ("psychic pain")
    - Somatic complaints
    - Preoccupation with the deceased
    - Emotional distance
    - Irritability
- Acute grief may last 4 to 6 weeks if there is no significant pathologic condition and there is some support.
    - Mourning is a lengthier process (1 to 2 years).
    - Bowlby described three stages to mourning:
        - *Protest*—disbelief, anger
        - *Despair*—anxiety, depression
        - *Detachment*—reorientation to life without the deceased

Table 7.2 *Pathologic Grief Reactions*	
Absence of grief	Dysfunctional denial
Extended grief	Dysfunctional hostility
Manic escape	Clinical depression

- A *grief disorder* refers to extended, delayed, or otherwise atypical grief reactions that result in distress or impaired functioning (Table 7.2).
- **Support and treatment**
    - With appropriate social support, most people succeed in resolving the experience of grieving without professional treatment.
    - Suppression of grief with medications can be counterproductive, but pathologic reactions can be addressed with both therapy and antidepressant medications, in addition to considerate and respectful care.

Multiple Choice Review Questions

1. Among adults over 65 years of age, which of the following is *true?*

   a. About 25% live in institutional settings.
   b. The poverty rate is higher than for younger adults.
   c. Average life expectancy is 8 years for men and 12 years for women.
   d. Being widowed is a more common experience for women than men.

2. Normal changes associated with aging include which of the following?

   a. Reduced time spent in deeper stages of sleep
   b. Reduced sensitivity to medications with anticholinergic effects
   c. Decrease in the size of the cerebral ventricles
   d. Narrowing of the cerebral sulci

3. Elderly adults are likely to complain of all *except* which of the following sexual problems?

   a. Reduction in vaginal lubrication
   b. Taking longer to achieve an erection
   c. Having a shorter refractory period
   d. Taking longer to reach orgasm

4. With regard to dementia, it is correct to say which of the followings?

   a. About half of all cases are reversible
   b. Dementia caused by reaction to medication is called pseudodementia.
   c. Dementia is equally common at age 65 years and age 85 years, but with different causes.
   d. About half of dementia cases are the result of Alzheimer's disease.

5. According to Elisabeth Kübler-Ross, patients who know they are dying do which of the following?

   a. Almost universally experience a period of clinical depression
   b. Commonly respond to the terminal prognosis with initial disbelief
   c. Do not usually respond to antidepressant medication when they become depressed
   d. Pass through a series of well-defined stages, leading finally to acceptance

PART 3
Behavioral Health

Chapter 8

The Doctor-Patient Relationship

CLINICAL COMMUNICATION

Roles and Expectations

- Physicians are expected to be *experts* having specialized knowledge and skills.

  - The *explicit* purpose of communication is to fulfill that role—*gathering information, informing the patient about diagnoses and treatments, implementing procedures.*
  - Communication also has *implicit* purposes, such as *expression of concern, reassurance,* and *development of the relationship.*

- Expectations of individual patients reflect their unique social, medical, and personal histories.

  - Patients react to specific events such as symptoms or diagnoses on the basis of individual personality and previous experience with physicians.
  - *Transference* occurs when a patient's attitudes and expectations are shaped by experiences with others (e.g., parents, family members).
    - Transference may be *positive* or *negative* and the patient is *unconscious* of it.
    - *Positive transference* can be dysfunctional if the physician is overly idealized or becomes the target of sexual attraction.
  - *Countertransference* occurs when physician attitudes and behavior are similarly influenced.

- *Optimal* relationships depend to some extent on the clinical situation.

  - A *passive* patient is more desirable for acute conditions and emergency care.
  - An *active* patient is more desirable for most other situations:
    - Management of chronic conditions
    - Preventive care
    - Aspects of care that emphasize life-style changes such as diet or exercise

- *Rapport* is a sense of understanding and trust.

  - Developing rapport may represent a real challenge when patients differ significantly from the physician in terms of education or social status.
  - Rapport influences the patient's perception of the physician's technical skills.

- Physicians with more positive "bedside manner" are viewed as more competent, even though competence is unrelated to bedside manner.
- Physicians perceived more warmly are less likely to be sued.

■ *Setting* can be influential if it is intimidating, hurried, and rushed or does not provide a minimal sense of privacy.

■ **Interviewing Techniques**

- *Open-ended questions* are the least restrictive and produce spontaneous information.
- *Direct questions* are useful for specific detail or when time is lacking or when the patient is limited in capacity to respond.
- *Facilitation* encourages clarification and elaboration ("Tell me more about that.").

  - *Silence* can be facilitative.
  - *Reflection* is another variant of facilitation ("You get headaches from any strong smell?").

- *Confrontation* asks the patient to resolve or respond to inconsistencies ("I know you don't like the medication, but you haven't been willing to really discuss alternatives like changing your diet and losing weight. What should we do?").
- *Validation* reassures the patient that symptoms or reactions are appropriate and understandable.
- *Recapitulation* restates the patient's information so that the patient can confirm the accuracy of the physician's understanding or make corrections.
- Expressions of *empathy* and *support* establish the physician's interest and concern and encourage the patient to be forthcoming.

ADHERENCE AND COMPLIANCE

Adherence and compliance are terms referring to the extent to which the patient follows medical advice. Studies show that *physicians tend to overestimate adherence.*

■ **Clinical Realities**

- Studies indicate that on average two thirds of hypertensive patients do not adhere to their medication regimens.
- As many as half of glaucoma patients are not adherent to treatment requirements despite the threat of blindness.
- Failure to complete a prescribed course of antibiotics is notoriously high.
- Research does not support the existence of a *noncompliant personality.*

■ **Factors Improving or Discouraging Adherence**

- The *quality of the doctor-patient relationship* is an important factor in determining adherence.
- Adherence is also influenced by characteristics of the *patient, illness,* and *treatment* (Box 8.1).

Box 8.1

FACTORS IN ADHERENCE

Factors Facilitating Adherence
Acute illness
Observable and uncomfortable symptoms
Clear and written instructions
Simple and time-limited regimen
Social support
Patient understanding of illness and treatment
Ensuring patient skills (e.g., ability to evaluate fat in foods, administer injections)
Specific follow-up study

Factors Discouraging Adherence
Chronic illness
Complex treatment regimen
Requirement of life-style changes
Asymptomatic medical condition
Lack of necessary skills
Unsupportive social environment
Lack of understanding of illness and treatment

Box 8.2

ELEMENTS OF THE HEALTH BELIEF MODEL

Patients believe that
    The illness has serious consequences
    They are personally at risk
    They can carry out the treatment
    The treatment will reduce the threat

EMOTION AND BEHAVIOR IN ILLNESS

■ Seeking Medical Care

- Whether a patient seeks care depends on how symptoms are perceived and whether symptoms disrupt activities.
- In the *health belief model,* patients seek care and follow instructions based on their beliefs about consequences and personal risk (Box 8.2).
- Most patients will make some effort to treat a problem on their own, possibly using "over-the-counter" medications, before seeking care.
- Persons with physical symptoms are twice as likely to seek medical care if they are experiencing stressful events or circumstances.

■ Responding to Patient Emotions

- "Good patients" are those who do not express more emotion than the physician believes is appropriate. Strong emotions may be perceived as time consuming and strain the physician's capacity for empathy.
- Emotionality can be difficult for physicians who have overly strong needs for distance or whose need to be nurtured is overly high.

■ The Sick Role and "Illness Behavior"

- The *sick role* refers to the *social* allowances and expectations that apply to an ill person (Box 8.3).

> **Box 8.3**
>
> ### ELEMENTS OF THE SICK ROLE
>
> The patient is not responsible for the illness and the need for care
> The patient is excused from normal responsibilities such as work
> The patient should be motivated to get well as quickly as possible
> The patient should seek appropriate and competent care

> **Box 8.4**
>
> ### ELEMENTS OF ILLNESS BEHAVIOR
>
> **Personal Meanings Attached to Illness**
> Challenge to be met
> Evidence of personal weakness
> Suspension of normal life
> Justification for strong emotional expression
> Punishment
> Becoming "damaged goods"
>
> **Coping Responses**
> Minimization and denial
> Avoidance
> Hyperattention and somatic preoccupation
> Refusal to accept limitations
> Capitulation or surrender
> Seeking information from other sources
> Pursuing "alternative" health care

- *Illness behavior* refers to how *individuals* respond to having an illness and the personal meaning they attach to it (Box 8.4).

■ *Placebo* effects are improvements in the patient's condition that are not attributed to specific treatment but rather to meeting *expectations* about effective treatment.

- The doctor-patient relationship and the patient's perception of care as effective are important mediators of these effects.
- Placebo effects are observed in both subjective aspects of illness and in objective physical indicators.
- Placebo effects influence both psychogenic and organic symptoms; a placebo effect is *not* an indication that symptoms are "all in the patient's head."
- There is considerable variability across studies, but on average about one third of patients will have a positive response to placebos.
- *Double-blind controlled studies* are used to account for placebo effects in evaluating efficacy of new treatments.

Chapter 8   The Doctor-Patient Relationship

MULTIPLE CHOICE REVIEW QUESTIONS

1. Transference is best described as which of the following?
   a. The building of satisfactory rapport between physician and patient
   b. The physician's pursuit of implicit goals in communication with a patient
   c. The patient's deliberate effort to control how he or she is perceived by the physician
   d. A patient's unconscious emotional reaction to the physician

2. A patient tells you following surgery that his physical therapy sessions leave him very sore for a while afterward, and he has refused his last session because he is afraid it is hurting him. Which response from you would illustrate validation?
   a. "You can't expect to get better if you don't cooperate with therapy."
   b. "Everyone is sore after physical therapy and often thinks that something is wrong."
   c. "So you are worried that the soreness is a sign that something is wrong?"
   d. "Where is the soreness, exactly, and how long does it last?"

3. We know from studies of compliance that which of the following statements is *true?*
   a. Physicians often overestimate the extent to which patients comply with treatment.
   b. Psychological screening tests can help identify patients with the noncompliant personality type.
   c. Compliance is primarily the result of effective physician behaviors.
   d. Compliance is more likely in patients who require ongoing treatment.

4. Factors discouraging compliance include all *except* which of the following?
   a. The patient's condition is asymptomatic.
   b. The patient has to make changes in diet and exercise habits.
   c. The patient's illness is acute.
   d. The treatment involves multiple medications.

5. Patients who exhibit a placebo response do which of the following?
   a. Report subjective improvement, but physical indicators are unchanged
   b. Usually have strongly psychosomatic symptoms
   c. Can show changes in both psychogenic and organic symptoms
   d. Consistently make up about 10% of those given a placebo.

Chapter 9

Stress

DEFINING STRESS

- Stress can be understood as a *stimulus*, a *response*, or an *interaction between person and environment* (Table 9.1).
  - Holmes and Rahe quantified "stress as stimulus" with the *Social Readjustment Rating Scale*.
  - Life events were given scores reflecting relative severity *(life-change units)*.
    - Death of a spouse is considered most stressful and is assigned a score of 100.
    - Both *positive* and *negative* events are included; negative ones are typically more severe.
  - Accumulating high numbers of life-change units has been associated with increased risk of illness.
  - The approach is criticized for failing to take into account *individual perceptions* of events and *coping resources*.
  - Others have related illness to the less profound but ongoing minor sources of demand in life ("hassles"); this approach has the same failings.
- The *general adaptation syndrome,* or GAS (Selye) is a description of a nonspecific pattern of *response*.
  - The three stages of the GAS are the *alarm, resistance,* and *exhaustion* stages.
  - Sustained resistance, even before exhaustion, takes a toll leading to *diseases of adaptation*.
- The effect of a stressor is mediated by perceptions or *appraisals*.
  - *Primary* appraisal is evaluation of the level of threat posed by a stressor.
  - *Secondary* appraisal is evaluation of coping skills and ability to use them effectively.

Table 9.1	*Stress Concepts*
Stimulus	External events that require adaptation such as life changes and "hassles"
Response	Internal events such as increased heart rate or endocrine changes
Interaction	Individual interpretation of and response to circumstances

Chapter 9 Stress

PHYSIOLOGY OF STRESS

Neuroendocrine Effects

- Activation of the *hypothalamus* by the *limbic system* stimulates release of *corticotropin-releasing factor (CRF)*.
- CRF stimulates pituitary release of *adrenocorticotropic hormone (ACTH)*.
- ACTH stimulates release of *corticosteroids* from the adrenal cortex.
- Stress promotes release of *catecholamines* from the *adrenal medulla* (Box 9.1).
- Stress is also associated with increased production of *endorphins*.

- Adverse effects on *immune functioning* are seen in various stressful circumstances including *divorce, death of a spouse,* and *academic pressure*.
- **Stress and Illness** There is evidence that stress is a contributing factor in a variety of conditions (Box 9.2).

MODIFIERS OF STRESS

Personality Factors (Table 9.2)

- The *type A behavior pattern* has been associated with increased risk of *myocardial infarction*. Studies show that it is specifically *hostility* that is associated with increased risk.
- *Psychologic hardiness* is associated with better resistance to stress. Benefits of hardiness are more reliably observed in men according to existing research.

Box 9.1

ADRENAL RESPONSES TO STRESS
Adrenal Cortex (Corticosteroids) Gluconeogenesis Increased insulin blood levels Reduced inflammatory responses Reduced antibody responses Increased gastric secretion **Adrenal Medulla (Catecholamines)** Increased heart rate Increased respiration Increased blood pressure Diaphoresis Decreased gastrointestinal motility

Box 9.2

STRESS AND INCREASED HEALTH RISKS
Coronary artery disease Hypertension Cancer Sudden cardiac death Obstetric complications Psychiatric disorders

Table 9.2	Type A Behavior and Hardiness
Type A Behavior	**Psychological Hardiness**
Impatience	Perception of stressful events as challenges
Time urgency	Firm sense of self
Hostility	Sense of meaning in life
Competitiveness	Feeling able to influence events and outcomes

■ **Coping Skills and Resources**
- The usefulness of coping strategies depends on the combination of person and situation.
    - *Problem-focused coping* involves an emphasis on changing the situation.
    - *Emotion-focused coping* involves an emphasis on changing how one reacts to the situation.
- Features of the physical environment such as noise, pollution, and overcrowding mediate the stressfulness of events and the capacity to respond effectively.
- Socioeconomic factors such as education and income mediate stress.
    - Poverty is itself a stressor.
    - Lower status is associated with negative differences in diet, exercise, and health habits and with less access to resources.
- Social isolation and lack of social support are strongly associated with morbidity and mortality risks.

Stress Management

■ *Relaxation techniques* include *progressive muscle relaxation, biofeedback,* and *meditation.*
- These techniques modify the stress response by *lowering levels of physiologic arousal* and *enhancing the person's sense of control.*
- No technique, including *hypnosis,* is specifically superior.

■ *Cognitive strategies* emphasize the way people think about events and consequences.
- *Cognitive reframing* involves selecting the most adaptive view of situations that can be interpreted in different ways.
- *Cognitive therapy* involves indentifying *maladaptive assumptions* and other dysfunctional thinking.

■ *Stress inoculation* involves preparation for stress through *skills training* and *behavioral rehearsal.*

■ *Exercise* is associated with better resistance to stress, reduced risk of depression, improved self-esteem, and better sleep.

■ Some persons need specific *skills training* in areas such as assertiveness, time management, parenting, and communication.

Chapter 9   Stress

STRESS AND MEDICAL CARE

- *Depression and anxiety* are common responses to illness as people respond to the threat illness represents and the associated disruptions of life.
  - Overt emotional expressions of distress include tearfulness, withdrawal, irritability, and anger.
  - Reactions may include *regressive behavior.*
  - Risk of adverse psychologic reaction is higher in patients with a prior history of psychiatric disorder and patients who lack social support.
- **Psychiatric Consultation**
  - Consultation may be required when patients are agitated, suicidal, or refusing treatment.
  - Mental health consultants can make a variety of contributions.
    - Diagnosing psychiatric disorders
    - Providing support to patients and families
    - Counseling staff and family members on how to respond to problematic reactions
    - Recommending specific problem solutions including medications
- **High-Risk Situations for Emotional Disturbance**
  - Life-threatening illnesses (e.g., AIDS)
  - Intensive care units
  - Surgery
  - Renal dialysis
  - Stroke
- *Support groups* are often helpful, as are procedures that enhance a patient's sense of control.

CHRONIC PAIN

Basic Concepts and Issues

- As much as 80% of physician visits are prompted by pain; more than 5 million Americans are disabled by chronic pain, half of them permanently.
- *Nociception* refers to the physical process of stimulation of specialized nerve fibers associated with pain.
  - *A delta (Aδ) fibers* transmit pricking and localized pain and prompt rapid withdrawal from the stimulus.
  - *C fibers* transmit dull and burning pains.
- The phenomenon of pain also includes the subjective unpleasant experience (unobservable) and observable *pain behavior.*
  - Pain behavior includes verbal (complaints) and nonverbal (guarded movements, facial expressions) components.
  - Pain behavior is both a *response* to unpleasant stimulation and an *operant* shaped by the environmental response.
  - When pain behavior is *reinforced* by attention or release from responsibility, the pain behaviors may be increased.
  - Complications may develop when healthy behavior is not reinforced.

Box 9.3

ANALYSIS OF PAIN BEHAVIOR
Evaluation of daily activities
Identifying antecedents of pain complaints and behaviors
Identifying consequences (environmental responses) of pain behaviors
Identifying maladaptive responses and behavior patterns
Evaluation for depression and other complicating psychiatric conditions

- These concepts apply much more directly to chronic pain.

■ *Assessment of chronic pain* should include careful physical examination, analysis of pain behavior, and evaluation of psychologic and physiologic responses.

- Physical examination may identify contributory phenomena that can be treated.
- *Analysis of pain behavior* involves collecting information about the situation and the patient's responses, usually through interview with patient and family members (Box 9.3).
- Psychologic tests such as the Minnesota multiphasic personality inventory and *pain questionnaires* are commonly used.

■ **Behavioral Management of Chronic Pain**

- *Contingency management* involves arranging the environment to maximize reinforcement of healthy behavior.
- *Biofeedback and relaxation techniques* emphasize control over physiologic responsiveness and emotional arousal.
- *Cognitive-behavioral treatment* alters the patient's interpretations of events (e.g., I hurt; therefore I should not be active) and encourages a sense of control.
- Combinations of environmental management, skills training, and promotion of optimal outlook have the most potential benefit.

Chapter 9 Stress

MULTIPLE CHOICE REVIEW QUESTIONS

1. The primary appraisal of a stressor is which of the following?
   a. An evaluation of one's coping skills and resources for dealing with the stressor
   b. A judgment about whether the stressor is subjectively negative or positive
   c. An evaluation of the level of threat posed by a stressor
   d. A judgment about whether one can cope with the stressor effectively

2. Which of the following are diseases of adaptation?
   a. Those that occur in the alarm phase of the General Adaptation Syndrome
   b. Those that result from sustained demands for resistance to stress
   c. Those that occur when the organism reaches the exhaustion phase of the General Adaptation Syndrome
   d. Those that are a response to ongoing "hassles" rather than life-changing events

3. With respect to the Type A behavior pattern which of the following is *true?*
   a. It appears that the feature of hostility is most strongly associated with risk of illness.
   b. Individuals who exhibit Type A behavior are at increased risk of cerebrovascular disease.
   c. Individuals who exhibit Type A behavior tend to view stressors as challenges.
   d. The link with myocardial infarction has been disproven.

4. All *except* which of the following statements about stress management techniques are correct?
   a. Cognitive reframing emphasizes finding the most useful interpretation of stressors.
   b. Hypnosis is significantly more effective than other relaxation techniques at reducing arousal.
   c. Exercise promotes resistance to stress and better sleep.
   d. Relaxation techniques contribute to an enhanced sense of control in the face of stressors.

5. Pain behavior is described by which of the following?
   a. It consists of nonverbal indicators that a person is experiencing pain.
   b. It refers to responses to pain that are not influenced by reinforcement.
   c. It consists of responses directly related to nociception.
   d. It may be influenced by an environment that encourages it.

Chapter 10

Substance Use

BASIC CONCEPTS

- *Substance-abuse disorders* are diagnosed when use of a substance leads to impaired physical or psychosocial funtioning.
- *Dependence* is defined by the occurrence of a *withdrawal* syndrome when use of the substance is discontinued.
  - Withdrawal symptoms may include *psychologic* phenomena such as anxiety, depression, and *cravings* for the substance.
  - Dependence is usually characterized also by *tolerance,* in which increasingly higher amounts of the substance are required to produce the desired effect.

ALCOHOL ABUSE

Basic Concepts
- Alcohol is used by about 67% of American adults.
  - About 75% of men and 60% of women use alcohol.
  - Consumption has declined in the last decade, except among teenagers and young adults.
- Beverage alcohol is *ethanol.*
  - *Alcohol dehydrogenase* converts ethanol to *acetaldehyde,* which is converted first to *acetic acid* and then to carbon dioxide and water by *aldehyde dehydrogenase.*
  - 90% of ethanol is metabolized in the liver at a rate of about ½ oz. per hour.
  - *Legal intoxication* is defined as a blood level of *100 mg/dl* in most jurisdictions.
  - Levels above *400 mg/dl* are associated with *surgical anesthesia* and can produce *coma* and *death* in nonalcoholics.
  - Although ethanol is a central nervous system *depressant,* initial effect may be *behavioral disinhibition.*
  - *Benzodiazepines* show cross-tolerance and cross-dependence with alcohol, which accounts for their effect in blunting the symptoms of alcohol withdrawal.

Alcoholism

- Patterns described as *type 1 and type 2* have been described, differing in age of onset, sex distribution, and correlates (Table 10.1).
- Assessment tools of use to the primary care physician include the *CAGE questionnaire* and the *Michigan Alcohol Screening Test (MAST).*
  - The *CAGE,* the name of which is a mnemonic code, consists of four questions dealing with attempting to Cut down, being Annoyed by criti-

Chapter 10 Substance Use

Table 10.1 *Type 1 and Type 2 Alcoholism*	
TYPE 1 (MILIEU LIMITED)	TYPE 2 (MALE LIMITED)
Older age of onset	Earlier age of onset
Both males and females	Much more often males
Drinking relieves dysphoria	Drinking induces euphoria
More loss of control	Less loss of control
More guilt but less sociopathy	Less guilt but more sociopathy

cism, feeling **G**uilty about drinking, and taking a morning drink (**E**ye-opener).

- Positive answers to three questions indicate alcoholism with 95% certainty.

- *Genetic factors* are well established in the pathogenesis of alcoholism.
    - Concordance is approximately 60% in monozygotic twins and 30% in dizygotic twins.
    - Scandinavian adoptions studies also show a strong genetic component.

- Alcoholism in women
    - Alcoholism has a later age of onset in women.
    - Differences in alcohol metabolism appear to place women at risk for adverse effects at lower doses and shorter duration of abuse.

Effects of Alcohol Abuse

- Alcohol abuse is associated with *liver disease, pancreatitis, gastrointestinal disorders,* and *cardiac disease.*

- Neuropathologic conditions include *alcohol hallucinosis, withdrawal delirium, amnestic disorder,* and *alcohol-induced persisting dementia.*

- Other consequences
    - Alcohol is a factor in as much as *a third of all general hospital admissions.*
    - It contributes to half or more of fatal motor vehicle accidents, homicides, and drownings.
    - Alcoholics have a *mortality 2.5 times higher than that of the general population.*

- *Fetal alcohol syndrome,* resulting from maternal alcohol consumption in pregnancy, is seen in about 50,000 births annually. The syndrome is characterized by *facial abnormalities, growth deficiencies, mental retardation,* and *behavior problems.*

Treatment of Alcohol Problems

- *Detoxification* can require management of a variety of *withdrawal* symptoms (Table 10.2).
    - *Benzodiazepines* may be used to mitigate the potential rigors of detoxification.
    - *Thiamine* is routinely administered to prevent development of *Wernicke-Korsakoff syndrome.*

- *Alcoholics Anonymous* (AA) is a voluntary self-help program built around

Table 10.2	Alcohol Withdrawal
Simple withdrawal	Tremors, tachycardia, diaphoresis, nausea, malaise
Motor seizures	
Withdrawal delirium (delirium tremens)	Delirium, autonomic hyperactivity, tremulousness, hallucinations, agitation
Hallucinosis	

the support of fellow "recovering" alcoholics. Associated groups include *Alanon* for the alcoholic's significant others and *Alateen* for youngsters, usually children of alcoholics.

- *Inpatient programs* commonly involve a 4-week plan with detoxification followed by therapy and rehabilitation.
- *Disulfiram* (Antabuse) discourages drinking by causing an intensely unpleasant physical response to alcohol.

TOBACCO USE

- About *30% of adults smoke cigarettes.*
  - Smoking has declined in the past two decades, *except among teenagers and African Americans* and in many other nations.
  - Smoking is more common among psychiatric patients.
  - Less than 10% of physicians now smoke.

Consequences of Smoking

- Nicotine produces addiction, and withdrawal is characterized by *headaches, irritability, weight gain,* and *impaired concentration.*
- Smoking is associated with increased risk of *lung cancer, chronic obstructive pulmonary disease (COPD), heart disease,* and *stroke.*
  - Pregnant women who smoke have an increased risk of *spontaneous abortion* and *low birth weight infants.*
  - Children of smokers have more frequent *respiratory tract infections* and *middle ear disease.*
  - Smoking accounts for more than 400,000 premature deaths annually and more than $50 billion in direct and indirect costs.

Smoking Cessation

- Although virtually all physicians identify smoking as a major health problem, studies suggest that most do not believe they know how to counsel smokers effectively.
  - Less than 75% routinely advise patients to quit.
  - Young males and light smokers are counseled less often than others.
- *Withdrawal* symptoms include irritability, weight gain, headaches, and cravings.
  - Symptoms peak 1 or 2 days after cessation.
  - Patients are often worried about gaining weight, but the average gain is only about 5 pounds.

- A wide range of cessation strategies exist with success rates at 1 year of follow-up study ranging from 20% to 40%.
  - Multicomponent strategies including specific attention to relapse prevention appear to be most effective.
  - Follow-up examination at less than a year leads to overestimation of success.

■ *Smokeless tobacco* does not involve lung disease risks but is addictive and does pose risks of cancer of the oral cavity.

OTHER DRUGS

■ *Opioids* include substances directly derived from the opium poppy (morphine), substances derived synthetically from naturally occurring opiates (heroin and codeine), and synthetic narcotics.
- Effects include *analgesia, euphoria,* and *sedation.*
  - Overdose is characterized by a triad of *coma, pinpoint pupils,* and *respiratory depression.*
  - Death from overdose is usually attributable to *respiratory arrest.*
  - Effects can be blocked and overdose treated by antagonists such as *naloxone* and *naltrexone.*
- Opioids may be taken orally or "snorted" and may be injected either intravenously or subcutaneously; injection involves the risk of HIV infection.
- **Addiction**
  - Heroin is particularly potent and produces a rapid euphoric effect and is the opiate most commonly associated with dependence.
  - Most opiate-dependent persons have an additional psychiatric diagnosis, most commonly depression, antisocial personality, and alcohol-use disorders.
  - About 15% of opiate-dependent persons attempt suicide.
- Opioid withdrawal is unpleasant but rarely fatal in the absence of other physical illness (Table 10.3).
  - Withdrawal reflects rebound hyperactivity of adrenergic neurons, which are less active with opioid use.
  - Withdrawal can be treated with *clonidine,* an adrenergic agonist.
- *Methadone* is a synthetic opioid used in the treatment of heroin addiction.
  - Physical effects are less pronounced but this permits better social functioning.
  - Taken orally, it does not pose risks found with injection.

■ **Sedatives, Hypnotics, and Anxiolytics (Table 10.4)**
- *Barbiturates* have been used as *hypnotics and anxiolytics,* but problems have largely limited them to be used as *anesthetics and anticonvulsants.*
  - Tolerance develops quickly, and effective dose can come very close to lethal dose, increasing the risk of accidental overdose.
  - Overdose produces death from *respiratory depression,* and dangers are greatly enhanced when they are combined with alcohol.

Table 10.3 Opioid Withdrawal

Stage	Duration of Symptoms	
Peak	48-72 hours	
Resolve	7-10 days	
Symptoms	Nausea	Stomach cramps
	Lacrimation	Rhinorrhea
	Diaphoresis	Irritability
	Insomnia	Anorexia
	Yawning	Piloerection

Table 10.4 Anxiolytics and Hypnotics

Barbiturates	Benzodiazepines
Amobarbital	Alprazolam
Pentobarbital	Chlordiazepoxide
Phenobarbital	Diazepam
Secobarbital	Flurazepam

- Withdrawal involves risk of *grand mal seizures* and *withdrawal delirium* (Box 10.1).
- *Benzodiazepines* are used as anxiolytics, hypnotics, anticonvulsants, and muscle relaxants.
  - They are safer than barbiturates and have largely replaced them as anxiolytics and hypnotics.
  - Withdrawal from them is less severe, but symptoms are similar to those observed with barbiturates.
  - They produce less danger of respiratory depression in the absence of concomitant alcohol use.
  - They rapidly became widely prescribed after initial introduction in 1960s because risks of dependence were not recognized.
  - Prescriptions have decreased because of regulation and recognition of risks.

Stimulants

- *Amphetamines* are used primarily in the treatment of attention deficit hyperactivity disorder and narcolepsy.
  - Effects include euphoria, accelerated activity and increased energy, and anorexia.
  - Intoxication may be followed by a "crash," characterized by depression, anxiety, headaches, and fatigue.
  - Adverse effects include cerebrovascular, gastrointestinal, and cardiac effects and hypertension.
  - Sustained use may produce delirium or a psychosis characterized by paranoia, hyperactivity, hypersexuality, and visual hallucinations.
  - Classic amphetamines such as methamphetamine, dextroamphetamine, and methylphenidate affect dopaminergic systems.
  - "Designer" amphetamines such as *MDMA* ("ecstacy") influence serotonergic systems as well and produce additional hallucinogenic effects.

Box 10.1

BARBITURATE AND BENZODIAZEPINE WITHDRAWAL

Anxiety and agitation
Insomnia
Tremor and hyperreflexia
Diaphoresis and fever
Delirium
Seizures
Cardiovascular collapse

- "Ice" is a purified form of methamphetamine.
- *Caffeine* is a stimulant found in coffee, tea, cola drinks, and over-the-counter medications including stimulants and diet aids.
  - Effects include increased alertness and can include elevated blood pressure, tachycardia, insomnia, restlessness, increased gastric secretion, and diuresis.
  - Tolerance develops, and withdrawal is commonly characterized by headaches, fatigue, and mild depressive symptoms.
- *Over-the-counter stimulants* include ephedrine and propranolamine, found in nasal decongestants, and phenylpropranolamine, found in appetite suppressants.
  - Effects are less pronounced than in classic amphetamines, but they are subject to abuse because of ready availability.
  - In high doses they may provoke serious hypertension.
- Cocaine
  - It is available in powder usually inhaled or injected and in the form of "crack" or "rock," which is smoked.
  - The desirable effect is an intense euphoria lasting 30 to 60 minutes and followed by a "crash" of depression and fatigue similar to that observed with amphetamines.
  - Cocaine has pronounced addictive qualities especially in the crack form.
  - Cocaine has a variety of adverse effects, including potential for major medical complications (Table 10.5).
  - Withdrawal is characterized by intense cravings that can last for months.
  - It may produce an intoxication delirium and a psychotic disorder characterized by paranoid delusions and hallucinations.

Marijuana

- It is usually smoked and occasionally ingested in baked goods.
- The active ingredient is *tetrahydrocannabinol (THC)*.
- It produces mild euphoria, feelings of relaxation, somnolence, conjunctival injection, increased appetite lasting as much as 2 to 3 hours.
- It is associated on rare occasions with psychotic symptoms and somewhat more commonly with brief transient paranoid feelings.
- Risks of lung disease are associated with smoking it.
- An *amotivational syndrome* has been described, characterized by apathy and generally reduced activity.

Table 10.5 Adverse Effects of Cocaine	
ADVERSE EFFECTS	MAJOR MEDICAL COMPLICATIONS
Insomnia Poor concentration Irritability Weight loss Impaired judgment Symptoms of mania	Myocardial infarctions Cardiac arrhythmias Cerebrovascular accidents Seizures

- **Hallucinogens**
  - Examples
    - The best known hallucinogens are *LSD (lysergic acid diethylamide)* and *PCP (phencyclidine, or "angel dust")*.
    - Others include *psilocybin,* found in certain mushrooms, and *mescaline,* found in peyote.
  - *Effects* include alterations in perception such as hallucinations and spatial and temporal distortions.
    - Some users may have intensely negative experiences (*"bad trips"*) characterized by severe anxiety reactions.
    - Some users will have a return of the hallucinogenic experience in the absence of the drug *(flashbacks),* weeks or months later.
    - Hallucinogens are not associated with tolerance or withdrawal syndromes.
    - *PCP* is unique in its potential for producing episodes of *violent and combative behavior* and *long-term impairments in memory and attention.*
  - Physical effects and phenomena
    - Hallucinogens may produce tachycardia, tremor, palpitations, and diaphoresis.
    - PCP can induce seizures and can be fatal in overdose.
    - Effects of most hallucinogens last from 6 to 12 hours, but the effects of PCP may persist for several days.
- **Inhalants**
  - Substances of abuse include *solvents, adhesives (glues), and fuels.*
    - Active components (e.g., benzene, acetone, trichloroethylene) are well established as carcinogens and neurotoxins.
    - Substances are inhaled from plastic bags or soaked rags.
  - The "desired" effects include euphoria and "floating" sensations.
    - Other psychologic effects may include anxiety, hallucinations, emotional lability, and irritability.
    - Neurologic effects include ataxia, slurred speech, impairment of attention and memory, and headaches.
  - Serious adverse effects include potential for cardiovascular, pulmonary, and gastrointestinal disorders, and risk of hepatic, renal, and brain damage.
    - Neurotoxicity and hypoxia may contribute to irreversible dementia.
    - Death may result from cardiac arrhythmia or respiratory depression.
  - Tolerance develops, and withdrawal can be characterized by nausea, anxiety, tremulousness, diaphoresis, irritability, and insomnia.

Chapter 10   Substance Use

MULTIPLE CHOICE REVIEW QUESTIONS

1. When applied to substance use disorders, the concept of tolerance refers to which of the following?
   a. The emergence of a physiological withdrawal syndrome when use of the substance is discontinued
   b. The willlingness of an abuser to endure the physical and psychosocial consequences of continuing to use the substance
   c. The appearance of cravings and psychological distress when use of the substance is discontinued
   d. The need for increasingly greater amounts of the substance to achieve the desired effect

2. Which of the following statements about physicians and smoking is *correct?*
   a. About 20% of physicians smoke.
   b. More than 90% of physicians routinely advise smoking patients to quit.
   c. Most physicians feel well-equipped to assist patients in quitting.
   d. Physicians are less likely to counsel light smokers and young males to quit smoking.

3. Presenting symptoms of coma, pinpoint pupils, and respiratory depression indicate which of the following?
   a. Alcohol-induced delirium
   b. Severe alcohol withdrawal syndrome
   c. Opioid overdose
   d. An attempt to enhance the effects of opioids with naltrexone

4. Prolonged use of amphetamines can lead to which of the following?
   a. Delirium tremens
   b. Reduced alertness, blood pressure, and heart rate
   c. Delusions, paranoia, and visual hallucinations
   d. Decreased tolerance

5. A unique phenomenon found in use of hallucinogens is which of the following?
   a. The occurrence of "flashbacks" in some users in the absence of use
   b. The occurrence of an "amotivational syndrome" in many users
   c. The occurrence of paranoia in some users
   d. The occurrence of an intoxication delirium

Chapter 11

Sleep

NORMAL SLEEP

- *Sleep architecture* refers to the characteristic structure of sleep, divided into *non-REM sleep* (divided into several stages) and *REM sleep,* which is characterized by *rapid eye movements.*
  - Sleep stages are defined by characteristic *electroencephalogram (EEG) patterns.*
  - In addition to recording from the scalp, electrodes are placed to record *eye movements (electrooculography)* and *muscle tone (electromyography).*
- *Non-REM sleep* is so called because of the absence of rapid eye movements and has four stages.
  - The stages involve progressively deeper sleep in terms of requiring more sound stimulation to waken.
    - *Stage 1* is a brief transition period characterized by the disappearance of *alpha waves.*
    - *Stage 2* is characterized by *sleep spindles* and *K complexes.*
    - *Stages 3* and *4* are the stages of *deep sleep* characterized by increasing dominance of *delta waves.*
  - The stages cycle from 1 to 4 and then reverse in a pattern that takes about 90 minutes and is called the *ultradian rhythm.*
  - The average adult spends about *75% of the night in non-REM sleep.*
  - Somniloquy ('sleep-talking') occurs during non-REM sleep and is unrelated to dreaming.
- *REM sleep* is especially characterized by dreaming.
  - Physiologic features include *penile* and *clitoral erection* and *muscle atonia* (Box 11.1).
  - *Nocturnal emission (ejaculation)* in men reflects autonomic arousal and is not necessarily associated with sexual dream content.
- **Age-related Changes in Sleep Architecture**
  - Infants spend as much as 50% of sleep time in REM sleep; the proportion gradually declines to about 25% in young adults and declines further in older adults.
  - The deeper slow-wave sleep of stages 3 and 4 also diminishes in old age and may be virtually absent from the sixth decade.

SLEEP DISORDERS

- *Insomnia* consists of difficulty initiating or maintaining sleep.
  - As much as a third of the general population complains of poor sleep, and half describe the problem as serious.

Chapter 11 Sleep

Box 11.1

FEATURES OF REM SLEEP
Dreaming
Penile and clitoral engorgement
Nocturnal emissions
Muscle atonia
Increased pulse and respiration
Increased blood pressure

Box 11.2

CAUSES OF INSOMNIA
Systemic illness
Pain
Endocrine abnormalities
Medications
Substance use and withdrawal
Psychiatric disorders
Conditioned poor sleep
Shift work and schedule changes

- *Causal factors* include a variety of medical conditions and factors (Box 11.2).
    - *Conditioned poor sleep* can develop after an episode of illness or trauma disrupts sleep, and anxiety about being able to sleep becomes disruptive.
    - *Personality factors* associated with insomnia include *repressive coping style, inhibition of anger,* and *obsessional worry.*
    - *Periodic leg movement disorder* involves muscle contractions primarily occurring in lighter sleep and later in the night.
        - It may be accompanied by unpleasant dysesthesia in the muscles and irresistible urges to move that delay onset of sleep.
        - Prevalence increases after 30 years of age and may reach 50% among the elderly.

- Treatment options
    - *Medication* is best used very briefly, since even sleep medications *(hypnotics)* involve complications including *tolerance*.
        - *Benzodiazepines* reduce delta sleep.
        - *Antidepressants* reduce REM sleep.
    - *Behavioral techniques* (Table 11.1)
    - *Sleep hygiene* involves optimal behavioral and environmental practices.
        - Regular schedule
        - Daytime exercise
        - Quiet, dark, cool room
        - "Winding-down" time before bed
        - Elimination of caffeine and alcohol

Table 11.1 Behavioral Treatment of Insomnia	
Relaxation	Muscle relaxation, meditation
Sleep restriction	Bedtime limited to average sleep time of previous 2 weeks and gradually increased
Stimulus control	Out of bed for 20 minutes if not asleep in 20 minutes, and repeated until sleep onset is within 20 minutes
Thought stopping	Identification and disruption of thoughts associated with delayed sleep

Disorders of Excessive Somnolence

- Excessive daytime sleepiness caused by insufficient sleep is addressed by increasing sleep but may occur with normal amounts of sleep and then reflects other problems. Risks include accidents, poor job and school performance, and medical complications associated with specific disorders.
- *Sleep apnea* consists of episodic interruption of breathing with consequent hypoxemia.
  - It occurs in 1% to 5% of the population.
  - *Obstructive apnea* involves airway obstruction, and respiratory effort is signaled by *snoring*.
  - In *central apnea* there is no respiratory effort; there may be a mixture of central and obstructive features.
  - Sleep is *excessively light* and characterized by *multiple short awakenings*.
  - Apnea is associated with *males* (85%) and anatomic features including *obesity, jowls,* and *low-hanging soft palate*.
  - *Sustained hypertension* occurs in 50% of cases and *cardiac irregularities* in 40%.
  - Treatment may be behavioral, mechanical, or surgical.
- *Narcolepsy* is characterized by daytime sleep episodes that are irresistible.
  - Onset is within 5 minutes: episodes last 15 to 60 minutes.
  - Prevalence is estimated at 0.05%.
  - Associated symptoms include *cataplexy, sleep paralysis,* and *hypnogogic or hypnopompic hallucinations* (Table 11.2).
  - REM sleep is observed within a few minutes of sleep onset.
  - Treatment involves *stimulant medication*.

Parasomnias

- *Sleepwalking,* or *somnabulism,* is seen in deep sleep and typically lasts about 10 minutes.
  - It occurs in *15% to 30% of children*.
  - It persists into puberty in about 3% and into adulthood in about 1%.
  - It has a *family history correlation* in about 80% of cases.

Table 11.2 *Features Associated With Narcolepsy*	
Cataplexy	Abrupt loss of muscle tone that may be unpredictable but also occurs with strong expression of emotion
Hypnogogic or hypnopompic hallucinations	Dream features intruding into waking consciousness at sleep onset or sleep ending
Sleep paralysis	Atonia intruding into waking consciousness at sleep onset or sleep ending

- Although many obstacles are avoided, there is risk of injury from falls or collisions.
- *Sleep terrors* involve sudden waking with rapid pulse, hyperventilation, and possible screaming.
    - It does not occur in REM sleep and is *not associated with dreaming.*
    - It is seen in *5% to 10% of school-aged children,* tends to disappear in puberty, and may persist into adulthood in 1% of cases.
    - The child is not fully conscious and is not readily consoled or comforted.
    - The child usually falls quickly back into deep sleep.

Sleep and Health

- Both *long* (10 hours or more) and *short* (4 hours or less) sleepers *have higher mortalities. Sleep deprivation* is associated with dulled intellect, impaired concentration, moodiness, decreased sexual interest and response, higher accident rates, and impaired school or job performance. Extremes of sleep deprivation result in death in lab animals.

Multiple Choice Review Questions

1. Which of the following is *true* in the normal development of sleep architecture?
   a. Children do not exhibit REM sleep until about age 2 years.
   b. The amount of REM sleep gradually increases from childhood into adulthood.
   c. The amount of slow-wave, deeper sleep diminishes in old age.
   d. Slow-wave and REM sleep gradually increase in old age.

2. REM sleep is characterized by all *except* which of the following?
   a. Somniloquy (sleep talking)
   b. Dreaming
   c. Muscle atonia
   d. Penile and clitoral erection

3. Which of the following is *true* of sleep apnea?
   a. It occurs in as much as 15% of the population.
   b. It is associated with thin body and neck and high palate.
   c. It can result in dangerously low blood pressure.
   d. It is commonly signaled by a problem with very heavy snoring.

4. Which of the following best describes patients with narcolepsy?
   a. They are usually treated with sedative medications to make them sleep more deeply at night.
   b. They go into REM sleep within minutes of falling asleep.
   c. They typically have sleep episodes lasting 2 to 3 hours.
   d. They can have episodes of acute hypertonia and autonomic arousal, especially at times of strong emotion.

5. Which of the following is *true* of children with sleep terrors?
   a. They will usually have symptoms persisting into adulthood.
   b. They usually report intense and vivid dreams when awakened.
   c. They do not fully wake from an episode and are difficult to soothe.
   d. They are unable to go back to sleep for several hours after an episode.

Part 4

Psychopathology

Chapter 12

Mood Disorders

BASIC CONCEPTS

- *Mood* refers to a person's subjective emotional state, whereas *affect* refers to the external and behavioral manifestations of emotional state. Disorders reflect episodes of abnormal mood resulting in subjective distress and functional impairment (Table 12.1).
  - *Manic episodes* are characterized by elevated mood and activity levels.
  - *Hypomania* involves less dramatic elevation of mood and lesser levels of impairment.
  - *Depressive episodes* are characterized by dysphoric mood and diminished activity levels.

DSM-IV Disorders (Table 12.2)

- Depressive disorders include *major depression* and *dysthymia.*
  - Major depression is specified as representing a single episode or recurrence.
  - Severe depression is specified as being with or without psychotic features.
- Disorders incorporating episodes of mania or hypomania include the *bipolar disorders (I and II)* and *cyclothymia.*

Epidemiology

- Lifetime prevalence is many times greater for depression (10% to 20%) than for bipolar disorder (1%).

Table 12.1 *Features of Manic and Depressive Episodes*

MANIC	DEPRESSIVE
Euphoric mood	Sad or depressed mood
Grandiosity	Feelings of guilt or worthlessness
Decreased need for sleep	Insomnia or hypersomnia
Pressured speech	Fatigue or lack of energy
Distractibility	Poor concentration
Flight of ideas	Thoughts of death or suicide
Increased activity	Psychomotor retardation
Poor judgment	Indecisiveness

Table 12.2 *DSM-IV Disorders*	
Disorder	**Description**
Major depressive disorder	Meets criteria for depressive episode, may be recurrent or single episode
Dysthymia	Chronic but relatively mild depression
Bipolar I disorder	Has had at least one manic or mixed episode
Bipolar II disorder	Hypomanic episodes but never had full manic episode
Cyclothymia	Swings between depressed mood and hypomania without meeting full criteria for either manic or depressed episodes

Diagnostic and Statistucal Manual, edition 4, American Psychiatric Association, 1994.

- Mean age of onset is earlier for bipolar disorder (30 years of age) than for major depression (40 years of age).
- Depression is more common among women (about 2:1), whereas bipolar disorder is more equally distributed.

Clinical Features

- ■ *Psychotic symptoms* may be observed in both depression and bipolar disorder. Delusions and hallucinations are said to be *mood congruent* when the content is consistent with the emotional themes of depressed or elevated mood.
- ■ Variations are seen in how depression is presented.
  - Initial complaints may emphasize somatic complaints such as fatigue and sleep disturbance.
  - *Masked depression* is characterized by the absence of complaints of sadness and depressed mood.
  - *Melancholia* is characterized by
    - Pervasive loss of interest and pleasure and an inability to respond to pleasurable stimuli *(anhedonia)*
    - Prominent *vegetative* symptoms (e.g., disturbance in sleep and appetite)
    - Distinctively depressed mood and strong guilt feelings
  - Depression in the elderly, with diminished activity and complaints of poor memory, may on occasion look like dementia *(pseudodementia)*
- ■ **Course and Prognosis**
  - Depression
    - Most *untreated episodes* will resolve in about *6 months;* typical duration is reduced to *3 months with treatment.*
    - At least half of patients who have a depressive episode will have a second episode at some time.
    - About 20% of patients will develop chronic problems.

Chapter 12   Mood Disorders

- Bipolar disorder
  - Patients who have a manic episode typically have had previous episodes of depression.
  - Episodes have relatively abrupt onset and last from a few days to a few months.
  - Episodes end abruptly and are often followed by an episode of depression.
  - Prognosis is less favorable than for depression; risk of recurrence is high and exacerbated by common problems with medication compliance.
  - Chronic problems develop in about 30%.

- **Suicide**
  - About 15% of patients hospitalized for depression will die by suicide.
  - The likelihood of suicidal ideation and attempts will generally increase as the severity of depression increases.
  - Patients who are severely depressed may be unable to generate the energy to make a suicide attempt and may be at greater risk early in recovery when there is some increase in energy and initiative.
  - Risk is higher among those over 40 years of age, divorced or living alone, with a history of substance abuse, and with a history of previous attempts.
  - Females attempt suicide more often, but males more often succeed.

ETIOLOGY

- Genetic contributions are indicated from family risk, twin, and adoption studies.
  - **Dysregulation in virtually all** *neurotransmitter* systems has been implicated in mood disorders.
    - Antidepressants act to *inhibit reuptake of norepinephrine* or *serotonin*.
    - Abnormal levels of monoamine metabolites are associated with mood disorders (Table 12.3).
    - *Dopamine* activity appears to be lower during depressive episodes and elevated in mania.
    - *Acetylcholine* activity is also implicated.
      - Cholinergic activity is inversely related to monoamine activity; decreased monoamine activity is associated with greater cholinergic activity.
      - Cholinergic agonists can induce depressive symptoms.

Table 12.3 *Monoamine Metabolites in Depression*

NEUROTRANSMITTER	METABOLITE	MEASUREMENT
Serotonin	5-Hydroxyindoleacetic acid (5-HIAA)	Decreased cerebrospinal fluid (CSF) concentration
Norepinephrine	3-Methoxy-4-hydroxy-phenylglycol (MHPG)	Decreased in CSF and urine

- ### Psychosocial Factors
    - Stressful life events serve at least as precipitants among predisposed individuals.
    - Some studies have suggested that early life stressors such as loss of a parent in the first decade may create vulnerability to depression.
    - The causal role of depressive thinking patterns is ambiguous, but negative thinking (pessimism, hopelessness, negative interpretations of events, unrealistic beliefs) can contribute to the sustaining of a depressed mood. It is postulated that some individuals develop a *learned helplessness* in which they no longer can act to resolve problems.
    - The psychosocial disruptions of mood disorders create additional stresses.

- ### Drugs, Medications, and Nonpsychiatric Illness
    - Depression is common among abusers of central nervous system depressants such as alcohol and opiates. Stimulants such as methamphetamine can produce symptoms similar to mania; depressive symptoms often appear in the aftermath of stimulant use.
    - Prescription medications associated with depression include antihypertensives, tranquilizers, steroids, and oral contraceptives, among others.
    - Mood disturbance, particularly depression, is found in a variety of medical conditions (Box 12.1).

TREATMENT OF MOOD DISORDERS

- ### Antidepressant Medications
    - It typically takes *2 to 4 weeks* for benefits to become evident. Four to six weeks is considered an appropriate trial period for an antidepressant.
    - More recently developed medications such as the *selective serotonin reuptake inhibitors* (e.g., fluoxetine HCl, or Prozac) have rapidly become first line because of lesser side effects.
        - *Heterocyclic antidepressants* such as norpramine are efficacious but are noted for a variety of side effects (Box 12.2) and potential use in committing suicide.

Box 12.1

MEDICAL CONDITIONS ASSOCIATED WITH DEPRESSION
Hypothyroidism
Parathyroid disorders
Sleep apnea
Narcolepsy
Cardiopulmonary disease
Vitamin deficiencies
Multiple sclerosis
Cerebrovascular disease
Mononucleosis
Systemic lupus erythematosus
Parkinson's disease
Rheumatoid arthritis
Dementia
Pneumonia

Box 12.2

```
SIDE EFFECTS OF
HETEROCYCLIC ANTIDEPRESSANTS
```

Anticholinergic
Dry mouth
Constipation
Sweating
Tachycardia
Blurred vision

Other
Orthostatic hypotension
Cardiac arrhythmias
Sedation
Lower seizure threshold
Sexual dysfunction

- *Monoamine oxidase inhibitors* (MAOIs) are also efficacious, but adverse effects are similar to those observed with the heterocyclics and *require dietary restrictions*. Foods rich in tyramine can induce hypertensive crisis.

Mood Stabilizers and Antimania Agents

- *Lithium carbonate* is the first-line treatment for mania.
    - Antipsychotics may be temporarily necessary to achieve adequate behavioral control.
    - Blood levels must be routinely monitored to maintain a therapeutic level (0.9 to 1.4 mEq/L) and avoid toxicity.
    - A maintenance regimen is usually needed.
- *Anticonvulsants* such as carbamazepine or sodium valproate are used as alternatives or adjuncts.

Psychotherapy

- General applications
    - Psychotherapy establishes a supportive relationship, which may enhance medication compliance.
    - Patients may benefit from assistance in dealing with the psychosocial disruptions resulting from episodes of mood disorder.
    - Psychotherapy combined with medication is more effective than either treatment alone.
- *Cognitive approaches* emphasize adverse habits of thinking such as faulty assumptions and negative generalizations.
- *Behavioral techniques* may be employed in reinforcing increased activity and more positive verbalizations, developing problem-solving skills, and modification of the environment.

Electroconvulsive Therapy (ECT)

- ECT can be effective more quickly than antidepressants.
- Side effects are largely limited to transient memory loss.
- ECT is effective with both depression and mania.
- ECT is useful in severe disturbances and when medical conditions or allergies preclude usual medications.

Multiple Choice Review Questions

1. A patient exhibiting hypomania is likely to do which of the following?
   a. Display grandiose delusions and not sleep for several days at a time
   b. Exhibit manic symptoms but not necessarily require hospitalization
   c. Exhibit moderate to severe depression accompanied by restlessness and agitation
   d. Sleep and eat more than normal and display a very negative self-concept

2. When depression and bipolar disorder are compared, we find that which of the following is *true*?
   a. Depression is much more common than bipolar disorder.
   b. Major depression has a much earlier mean age of onset.
   c. Both disorders are more common in men than in women.
   d. Prognosis for bipolar disorder is more favorable than for depression.

3. Among patients presenting with a first episode of depression which of the following is *true*?
   a. About 85% will have a second episode at some point.
   b. Slightly more than half will develop a chronic mood disorder.
   c. Most will show abrupt onset and resolution of symptoms.
   d. Resolution of symptoms will typically take about 6 months if untreated.

4. All of the following statements about suicide are correct *except* which of the following?
   a. Likelihood of a suicide attempt generally increases with severity of depression.
   b. Males are more likely than females to attempt suicide.
   c. Risk of suicide is higher in adults over 40 years of age than in younger adults.
   d. Males are more likely than females to succeed in a suicide attempt.

5. In the pharmacologic treatment of mood disorders which of the following is *True*?
   a. Antidepressants act by reducing dopamine levels.
   b. Anticonvulsants are increasingly used as a first response to depressive episodes.
   c. Lithium is employed when depressive symptoms include psychotic features.
   d. Antidepressants typically take 2 to 4 weeks to produce improvement.

Chapter 13

Anxiety Disorders

BASIC CONCEPTS

- **Fear and Anxiety**
  - *Fear* is a response to an immediately present threat or danger.
  - *Anxiety* is a response to anticipated or potential threat, or it may occur without any awareness of a particular threat ("free-floating anxiety").
  - Fear and anxiety are not intrinsically abnormal or pathologic and are adaptive responses to many situations. Anxiety becomes a disorder when it results in excessive distress or impairment in functioning.

- **Signs and Symptoms of Anxiety (Box 13.1)**
  - Anxiety is a multifaceted response incorporating physiologic, affective, and cognitive components.
  - Manifestations of anxiety will vary, and an affected individual will not necessarily exhibit all features.

- **Biology of Anxiety**
  - Neurobiology
    - The actions of medications are suggestive of roles for neurotransmitters and associated structures (Table 13.1).
    - Abnormalities of the frontal and temporal cortex are associated with anxiety; frontal lesions can produce anxiety or an absence of anxiety.
    - Anxiety associated with pheochromocytoma (tumor of the adrenal medulla) is characterized by elevated cerebrospinal fluid levels of *vanillylmandelic acid (VMA)*, a metabolite of *norepinephrine*.
  - General medical conditions
    - Anxiety is encountered in a variety of neurologic, cardiovascular, pulmonary, and endocrine disorders.
    - Medications that can cause anxiety include stimulants, nonprescription decongestants and appetite suppressants, vasopressors, and anticholinergics.
    - Anxiety commonly accompanies withdrawal from alcohol, benzodiazepines, and other sedatives and hypnotics.

DSM-IV DISORDERS

- *Panic disorder* is characterized by discrete episodes of intense anxiety.
  - Symptoms include prominent autonomic arousal (diaphoresis, palpitations, trembling), shortness of breath, chest pain, paresthesias, fear that one is going crazy, and fear that one is dying.

Box 13.1

SIGNS AND SYMPTOMS OF ANXIETY
Nervousness
Trembling
Muscle tension
Sweating
Apprehensiveness
Dry mouth
Dizziness
Restlessness
Palpitations
Difficulty concentrating
Irritability
Insomnia
Fatigue
Hypervigilance
Exaggerated startle response
Numbness or tingling
Hyperventilation
Chest pain

Table 13.1 *Neurobiology of Anxiety*

MEDICATION	NEUROTRANSMITTER	STRUCTURE
Benzodiazepines	GABA	Limbic system
Antidepressants	Serotonin	Raphe nuclei
	Norepinephrine	Locus ceruleus
Antihistamines	Histamine	Hypothalamus

GABA, Gamma-aminobutyric acid.

- Onset is abrupt, and the typical episode lasts from 10 to 30 minutes.
- The initial attack is usually spontaneous, though patients will often have panic attacks in stressful situations as well.
- Patients will often present initially with fear they are having a heart attack.
- Panic attacks can be induced in susceptible persons by substances such as sodium lactate, carbon dioxide, sodium bicarbonate, yohimbine, and isoproterenol.
- Concordance rates among monozygotic twins are several times higher than for dizygotic twins.
- Complications include agoraphobia, other forms of conditioned fear and avoidance, depression (up to 50%), and alcohol abuse (up to 20%).

■ *Phobias* are fear and avoidance reactions that are well out of proportion to the potential threat represented by the stimulus.

● *Agoraphobia* (fear of open spaces) is a common complication of panic disorder.

- Patients begin to avoid being away from the safety of home for fear of having an attack in public or in the absence of help.
- Family relationships and vocational functioning may be seriously strained by the patient's refusal to leave home or go to public places unaccompanied.

- In severe cases, patients may be homebound for extended periods of time.
- Other phobias
  - *Social phobia* is characterized by fears of interacting with or being observed by others.
    - Fears focus on themes of being embarassed or humiliated.
    - Patients may have difficulty eating in restaurants, using public rest rooms, or performing routine tasks while someone is watching them.
    - Social phobia may become generalized, leading to avoidance of virtually all social situations.
  - *Specific phobias* usually involve objects or situations that could be threatening, but the fear and avoidance are disproportionate.
    - Typical phobias involve snakes, elevators, insects, heights, and flying.
    - Most phobias are isolated in their effects and do not prevent normal functioning in most domains of living.

Posttraumatic and Acute Stress Disorders

- *Posttraumatic stress disorder* develops after exposure to severe traumatizing experiences, such as actual or threatened death, serious injury, or loss of physical integrity (e.g., rape).
  - Symptoms involve *persistent reexperiencing of the trauma* (dreams, flashbacks), *avoidance of related stimuli*, *numbing of responsiveness* (loss of interest, restricted affect), and *persistent autonomic arousal* (hypervigilance, exaggerated startle reflex).
  - The most common precipitant for men is combat, whereas the most common precipitant for women is rape or physical assault.
  - Half or more of persons directly experiencing natural disasters (e.g., fires or earthquakes) will become symptomatic.
  - It can become chronic, with symptoms persisting for decades, and onset may be immediate or delayed for months or even years.
- *Acute stress disorder* is newly defined in the DSM-IV and involves symptoms and trauma similar to those defined in posttraumatic stress disorder. Symptoms last a *maximum of 4 weeks* and occur within 4 weeks of the precipitating event.

Generalized anxiety disorder is characterized by excessive worry about a variety of life circumstances and activities for at least 6 months.

- Depression and substance abuse are the most common complications, but as much as 25% of patients will develop panic disorder.
- The disorder is more common among females and tends to have onset before 30 years of age.
- Symptoms commonly include restlessness and tension, fatigue, irritability, difficulty concentrating, and insomnia.

Obsessive compulsive disorder is characterized by obsessions (persistent and intrusive thoughts or images) and *compulsions* (repetitive behaviors or "rituals").

- Clinical features
  - Both obsessions and compulsions are experienced as unwanted and cause anxiety or distress, and effort is made to suppress or ignore them.

- Compulsive behaviors (e.g., counting, handwashing, checking, and rechecking) often have a quasilogical link to a feared event or situation (e.g., exposure to germs), but the acts are an unrealistic response or are performed excessively.
- When patients resist or are prevented from performing the compulsive act, they feel a mounting anxiety that is discharged by the act.
- The time consumed by obsessional thoughts and compulsive acts is one of the sources of impairment.

● **Etiology and epidemiology**
  - Onset is typically in the teens or early twenties, and concordance is higher in monozygotic twins.
  - Neurobiologic models have emphasized abnormalities of the basal ganglia and prefrontal hyperactivity.
  - The neurotransmitter *serotonin* has been implicated, in part by the therapeutic benefits of medications that block serotonin reuptake.

TREATMENT OF ANXIETY DISORDERS

Antianxiety Agents
● *Benzodiazepines* are used primarily for acute anxiety and generalized anxiety disorder.
  - Alprazolam is useful with panic disorder.
  - Caution is required with these medications because of the potential for addiction or abuse.
● *Buspirone* is a nonbenzodiazepine anxiolytic that is similar to antidepressants in that benefits develop over a period of weeks. It is a useful alternative to benzodiazepines in generalized anxiety disorder and has little potential for abuse.
● *Beta-adrenergic receptor blockers,* such as propranolol, are used in controlling phobic anxiety and acute situational anxiety ("stage fright").

■ *Antidepressants,* including the cyclics, serotonin reuptake inhibitors, and monoamine oxidase inhibitors have applications in *panic disorder, agoraphobia, post-traumatic stress disorder,* and *obsessive-compulsive disorder. Clomipramine* is a cyclic antidepressant specifically indicated for obsessive compulsive disorder.

■ *Behavioral methods* have at least an adjunctive role in the treatment of most anxiety disorders.
● *Relaxation methods* can be used to control levels of arousal.
● *Systematic desensitization* and other *exposure therapies* are particularly useful in reducing anxiety provoked by specific stimuli or specific types of situations, as in phobias and panic disorder. Specific skills training (e.g., *assertiveness* and *social skills*) can be helpful with social phobia and other situational anxiety problems.

■ *Cognitive therapy* assists patients in controlling or redirecting anxiety-related thinking.
- *Thought stopping* and redirection assist patients in dealing with obsessional thoughts and the ruminative worry of generalized anxiety.
- Patients learn to manage "self-talk" by identifying anxiety-provoking thoughts and substituting more adaptive interpretations of events.

Chapter 13   Anxiety Disorders

MULTIPLE CHOICE REVIEW QUESTIONS

1. Which of the following is *true* in panic disorder?
   a. Agoraphobia is a common complication.
   b. Patients appear unusually resistant to depression.
   c. The first attack usually follows a traumatic stressor.
   d. The typical episode lasts 1 to 2 hours.

2. Mr. Jones reports that he is intensely anxious about using a public restroom, feeling a vague but strong sense that something embarrassing will happen. The most likely anxiety disorder is which of the following?
   a. Agoraphobia
   b. Generalized anxiety disorder
   c. Obsessive-compulsive disorder
   d. Social phobia

3. Which of the following is *true* in generalized anxiety disorder?
   a. The majority of patients will develop symptoms of panic disorder.
   b. Onset is usually after age 30 years.
   c. The disorder is more common among females.
   d. Depression is unusual.

4. Which of the following statements about obsessive-compulsive disorder is *correct*?
   a. The patient finds obsessional thoughts highly pleasant but cannot put them aside for other necessary activities.
   b. The patient rarely attempts to resist carrying out compulsive behaviors.
   c. Compulsive behaviors must have no reasonable or rational association with obsessional thoughts.
   d. Patients develop increasing feelings of anxiety if prevented from carrying out compulsive behaviors.

5. Which of the following is *true* in the pharmacologic treatment of anxiety?
   a. Buspirone is a preferred medication because of the more immediate onset of benefits.
   b. Antidepressants are useful in the treatment of panic disorder and agoraphobia.
   c. Beta-blockers such as propranolol are specifically useful for obsessive-compulsive disorder.
   d. Benzodiazepines are used with most disorders because of the low potential for abuse.

Chapter 14

Schizophrenia and Related Disorders

BASIC CONCEPTS

■ Terminology

- *Reality testing* refers to the ability to accurately perceive and interpret events and stimuli in the environment.
- *Psychosis* is a state in which reality testing is impaired and a person displays associated symptoms such as confusion, disorientation, delusions, or hallucinations.
- *Schizophrenia* is a major psychiatric disorder characterized by features including episodes of psychosis.
- *Thought disorder* is a term sometimes used to describe syndromes such as schizophrenia and the delusional disorders.

■ Epidemiology

- Peak age of onset is earlier for *males (15 to 25 years of age)* than for *females (25 to 35 years of age)*.
- Prevalence has been found to be associated with population density in very large cities and is disproportionately high in lower socioeconomic classes.
  - The association with social class is explained by the *downward-drift* hypothesis, which indicates that schizophrenics may be concentrated in lower classes because of their impairments.
  - Schizophrenics are estimated to make up between one third and two thirds of homeless persons.
- About half of psychiatric hospital beds are occupied by patients with schizophrenia.

SIGNS AND SYMPTOMS

■ *Premorbid and prodromal features* are observed, though usually identified in retrospect.

- *Premorbid* personality is commonly passive, introverted, and socially withdrawn. *Prodromal* changes in behavior are believed to reflect onset of disease and develop insidiously over a period of months or years.
  - Decline in social and academic or occupational functioning
  - Eccentric behaviors and interests
  - Multiple vague somatic complaints

Chapter 14 Schizophrenia and Related Disorders

General Features

- The *active phase* is characterized by prominent psychotic symptoms including disturbances in thinking, hallucinations, disorganized behavior, significant alterations in mood, and generalized impairment in functioning.
  - Symptoms of thought disorder can be categorized as disturbances in *thought form, content,* and *process* (Table 14.1).
  - *Hallucinations* are most often *auditory;* patients hear voices commanding, commenting, or criticizing.
  - Emotional disturbances include *flattened affect, inappropriate affect,* and *labile mood.*
- Negative and positive symptoms (Table 14.2)
  - *Negative* symptoms are characterized by absence or loss of normal behaviors and responses.
  - *Positive* symptoms are characterized by the presence of specific anomalous behaviors and responses.
- A *residual phase* follows active periods of illness.
  - Positive symptoms remit or diminish.
  - Negative symptoms often persist, with continuing impairment of functioning in life roles.

Table 14.1 Symptoms of Thought Disorder

Category	Example	Description
Form	Loose associations	Flow of ideas lacks logical connections
	Echolalia	Repetition of what others say
	Neologisms	Use of made-up words having meaning only to the patient
Content	Delusions	Fixed ideas that are bizarre or otherwise inconsistent with reality
Process	Flight of ideas	Thinking flits from one idea to another
	Blocking	Inability to formulate ideas or responses
	Inattention	Inability to sustain or focus attention

Table 14.2 Positive and Negative Symptoms of Schizophrenia

Positive	Negative
Hallucinations	Flattened affect
Delusions	Loss of motivation or interest
Bizarre actions	Thought blocking
Echolalia	Poverty of content in speech

- **Comorbidities**
    - Symptoms of depression are observed in 50% to 60% of patients with schizophrenia.
    - About half of patients make at least one suicide attempt, and up to 15% will succeed.
    - Schizophrenic patients are more likely than other psychiatric patients to smoke cigarettes.
    - Mortality from accidents and illnesses is higher in schizophrenic patients than that in the general population, and the majority have at least one concurrent physical illness.
    - About one third of patients will have alcohol-related disorders.
- **Subtypes of Schizophrenia (Table 14.3)** In addition to the residual syndrome, four specific suybtypes of schizophrenia are identified (*paranoid, catatonic, disorganized,* and *undifferentiated*).

COURSE AND PROGNOSIS

- The *prodromal* phase is followed by an active phase and then by the *residual phase*, which may be interrupted by further active phases or *acute exacerbations*.
    - The typical course is one of multiple exacerbations and remissions.
    - Remissions are characterized by incomplete return to base-line levels of functioning, and the base line tends to deteriorate after each acute episode.
    - About 50% of patients are severely impaired by the disorder for their entire adult lives.
    - Numbers vary with how a "good outcome" is defined, but about 20% of schizophrenic patients are able to function as integrated members of the community with some degree of participation in social roles.
    - *Positive* symptoms tend to *diminish* over the course of the disorder, and *negative* symptoms tend to become *more prominent.*
- **Good and Poor Prognostic Indicators (Table 14.4)**
    - The course of the illness over the first 5 years appears to predict the sort of course a patient will follow over time.
    - Many specific features are indicative of good or poor prognosis (Table 14.4).

Table 14.3 *Subtypes of Schizophrenia*

SUBTYPE	DESCRIPTION
Paranoid	Persecutory or grandiose delusions, auditory hallucinations, less prominent disorganization, and affective symptoms
Catatonic	Excitability or stupor, bizarre posturing, echolalia or echopraxia
Disorganized	Prominent disorganization of speech and behavior, flat or inappropriate affect, profound social impairment, and poor long-term functioning
Undifferentiated	Shows characteristic psychotic symptoms but does not meet criteria for other subtypes

Table 14.4	Good and Poor Prognostic Indicators
Good	**Poor**
Later or rapid onset	Early or insidious onset
More positive symptoms	More negative symptoms
Symptoms associated with precipitating events	Symptoms unrelated to precipitating stressors
No neurologic signs	Soft neurologic signs
Married or has friends	Unmarried or has few or no friends
History of employment	Poor or no work history
No family history of schizophrenia	Family history of schizophrenia

- Generally speaking, better functioning before the illness is predictive of a better outcome.

ETIOLOGY

- **Neurotransmitters**
  - *The dopamine hypothesis* relates schizophrenia to dopaminergic hyperactivity, reflecting the effectiveness of medications that block dopamine receptors.
  - Other neurotransmitters
    - *Norepinephrine* may be involved in paranoia and frequency of relapse.
    - *Serotonin* is implicated by the activities of newer or "atypical" antipsychotics and may be involved in suicidal behavior and abnormalities of mood.
    - *Gamma-aminobutyric acid* hypoactivity is associated with dopamine hyperactivity.

- **Neuroanatomic Correlates**
  - Computed tomographic and magnetic resonance imaging studies indicate *ventricular enlargement, cortical and cerebellar atrophy,* and *reduced limbic system volume.*
  - Studies of cerebral blood flow and metabolic activity indicate *hypoactivity of the frontal lobes* and *hyperactivity of the basal ganglia.*

- **Genetics**
  - Relatives of schizophrenic patients are at greater risk for the disorder than persons in the population are, and the risk is correlated with the closeness of the relationship.
  - Monozygotic twins have a concordance rate three times higher than dizygotic twins.

- *Psychosocial factors* are increasingly thought of as consequences rather than causes of the disorder. The diathesis-stress model indicates that stressful events may precipitate symptoms in vulnerable or predisposed individuals.

TREATMENT OF SCHIZOPHRENIA

- *Antipsychotic medications* are generally understood to work by blocking dopamine receptors.

- Traditional antipsychotics such as haloperidol (Haldol), thorazine, and thioridazine HCl (Mellaril) are about equal in efficacy.
- Some recently developed medications (e.g. clozapine and risperidone) may be more efficacious. Clozaril requires weekly blood monitoring because of a 1% to 2% risk of agranulocytosis.
- **Side effects**
    - *Anticholinergic effects* include dry mouth, blurred vision, sexual dysfunction, urinary retention, constipation, hyperthermia, and possible *delirium*.
    - *Extrapyramidal effects* include *bradykinesia* (slowed movement), *akathisia* (restlessness), *tremor,* and *dystonia* (rigidity and spasms).
    - *Tardive dyskinesia* is an extrapyramidal syndrome characterized by persistent involuntary movements of the mouth and tongue and choreoathetoid movements of the extremities.
        - It most often occurs in patients on antipsychotics for extended periods of time and can be irreversible.
        - Clozapine does not appear to be associated with tardive dyskinesia.

■ *Psychosocial interventions* provide emotional support, facilitate medication compliance, assist in structuring activity, and assist in dealing with the social complications of the disorder.

OTHER DISORDERS

■ *Schizoaffective* disorder is characterized by both schizophrenic symptoms and symptoms of mood disorder.
- Some psychotic symptoms must occur in the absence of significant mood symptoms.
- Course and outcome appear to fall in a middle ground between mood disorders and schizophrenia.
- *Subtypes* include *bipolar,* with outcome similar to bipolar disorder, and *depressed,* with outcome more similar to schizophrenia.
- *Poor prognosis* is associated with insidious and early onset, unremitting course, family history of schizophrenia, and absence of apparent precipitating events.

■ *Schizophreniform* disorder is characterized by symptoms similar to those of schizophrenia, but all symptoms (including residual) last no more than 6 months.

■ *Delusional disorder* is characterized by the presence of systematized *nonbizarre delusions,* whereas other aspects of personality and functioning are relatively well preserved.
- *Nonbizarre* delusions refer to events and situations that occur in real life, such as being followed, infected, or persecuted.
- Apart from the delusions, behavior is not generally odd or bizarre, and other symptoms of psychosis and thought disorder are not observed.
- Course tends to be chronic, but the disorder does not typically preclude adequate social functioning, except where the content of the delusion leads to conflict.

Chapter 14 Schizophrenia and Related Disorders

Multiple Choice Review Questions

1. Among patients with schizophrenia which of the following is *true*?
   a. Hallucinations are typically visual.
   b. Age of onset is earlier among females than among males.
   c. Negative symptoms include flattened affect.
   d. Symptoms of depression are unusual.

2. Which of the following is *true* in the course of schizophrenia?
   a. Following a period of active symptoms, patients typically do not completely return to their previous level of functioning.
   b. Most patients have only one period of active symptoms, followed by steady decline in functioning.
   c. Most patients exhibit a prodromal phase in which they have intense active symptoms of psychosis.
   d. About half of patients will be capable of fairly normal social participation, such as marriage and employment.

3. A more positive outcome would be indicated if which of the following were *true* for a patient with schizophrenia?
   a. A relatively young age at onset
   b. Prominent positive symptoms
   c. Symptoms without any precipitating stressor
   d. A pattern of primarily negative symptoms

4. Studies of schizophrenia have associated the disorder with which of the following?
   a. Dopamine hypoactivity
   b. Hypoactive frontal lobes
   c. Shrinkage of the cerebral ventricles
   d. Hypoactivity of the basal ganglia

5. Which of the following describes tardive dyskinesia?
   a. It is the movement disorder that identifies catatonic schizophrenia.
   b. It features a "waxy flexibility" and a tendency to mimic the actions of others.
   c. It is a side effect of long-term treatment with antipsychotic medication.
   d. It is a symptom of withdrawal from antipsychotic medications.

Chapter 15

Cognitive Disorders

BASIC CONCEPTS

- The primary feature of these disorders is disturbance in cognitive functions such as memory, attention, orientation, abstract thinking, and judgment.
- The cause is a known or presumed dysfunction of the central nervous system caused by
  - *Primary neuropathologic conditions* such as trauma, degenerative disease, neoplasm, and infection (e.g., encephalitis)
  - *Conditions secondary to systemic illnesses,* such as liver disease or endocrine disorders
  - *Substances* such as alcohol, medications, or poisons
- *Basic DSM-IV categories* of cognitive disorder include *delirium, dementia,* and *amnestic disorder.* Conditions that do not meet criteria for any of the specific cognitive disorders are classified as *cognitive disorder, not otherwise specified (NOS).*

DELIRIUM

- *Delirium* is a syndrome of generalized cognitive impairment accompanied by *disturbance of consciousness* and *impaired attention.*

Clinical Features

- *Onset is rapid* (hours to days), and severity tends to fluctuate over the course of the day.
  - Impairment is commonly observed to be more severe in the evening and late night hours, a pattern referred to as *sundowning.*
  - *Sleep-wake cycle* disturbances are seen.
- *Disturbance of consciousness* is manifest in reduced awareness of the environment and stupor or agitation.
- *Orientation* is typically impaired, particularly for time and place.
- Patients are unable to sustain, focus, or shift attention; other symptoms include *visual or auditory hallucinations,* confusion, anxiety, and incoherence.
- Patients with delirium incur a significant mortality risk; up to 50% will die within a year.

Risk Factors and Causes

- Delirium is seen in 10% to 15% of medical patients in the general hospital setting and up to one third of patients in surgical and intensive care units.
- Other risk factors include *old age, coexisting dementia,* and *previous history of delirium.*

Table 15.1 Selected Causes of Delirium	
SYSTEMIC	SUBSTANCES OR MEDICATIONS
Cardiovascular disease Endocrine disorders Electrolyte imbalance Liver disease Renal disease Systemic infection Vitamin deficiencies	Alcohol Anticholinergics Antihypertensives Opiates Phencyclidine (PCP) Poisons, neurotoxins Sedatives, hypnotics Steroids

- Delirium can be induced by a variety of neurologic conditions, including trauma, cerebrovascular disease, meningitis, and encephalitis.
- Delirium can also occur in a variety of systemic illnesses and with exposure to a variety of medications and substances (Table 15.1).

Clinical Management

- The primary task is treatment, when possible, of the causal condition.
- Environmental management emphasizes controlling the level of stimulation: providing a well-lit but quiet setting, consistent personnel, and prominent reminders of time, place, and situation.
- Agitation or disruptive behavior may require medication either with low dosages of antipsychotics or with benzodiazepines.

DEMENTIA

Dementia is characterized by impairment in cognition and in social and occupational functioning but lacks the disturbance in consciousness seen in delirium.

Clinical Features

- *Memory deficit* is a hallmark and is accompanied by other impairments such as *aphasia, apraxia,* or *agnosia.* Memory impairment typically involves recent rather than remote recall.
- The typical pattern is one of *insidious onset* and *progressive course,* but this will not be the case with some causes (e.g., dementia secondary to head injury).
- Behavioral changes may include initially subtle alterations of personality, decreased interests and activity, and reduced or labile emotionality.
- In *progressive conditions,* functioning becomes gradually more impaired in all domains. Patients become confused, lose the ability to carry out basic self-care activities, no longer recognize family and friends, and may develop psychotic symptoms including agitation and hallucinations.
  - Need for supervision escalates because of forgetfulness, impaired judgment, confusion, or wandering.
  - In patients who do not succumb to other conditions, coma and death eventually ensue.
- *Pseudodementia* refers to a condition primarily seen in *depressed* elderly persons whose symptoms may resemble dementia. Features differentiating dementia from pseudodementia are described in Table 15.2.

Table 15.2 Dementia versus Pseudodementia	
Dementia	**Pseudodementia**
Insidious onset	Relatively abrupt onset
Lack of concern	Expressions of distress
Deny or minimize cognitive dysfunction	Complaint of cognitive loss in some detail
Patient struggles with simple tasks	Patient gives minimal effort on simple tasks
Social skills and hygiene retained early on	Early loss of social skills and neglected hygiene
Absence of depressed mood or shallow affect	Pervasive depression or irritability
Less likelihood of prior psychiatric history	Greater likelihood of prior psychiatric history

Risk Factors and Causes

- *Alzheimer's disease* accounts for 50% to 60% of cases and is the most common of the degenerative dementias.
  - It is associated with degeneration of *cholinergic* neurons in the nucleus basalis of Meynert and with decreased brain levels of *choline acetyltransferase.*
  - Risk factors include increasing age, positive family history, female sex, lower levels of education and occupation, and history of head injury.
- *Multi-infarct dementia* is the second most common of the dementias, accounting for about 10% of cases.
  - The course is typically *stepwise,* with a drop in the level of functioning after successive episodes of cerebral infarction.
  - *The onset is usually more abrupt* than that in Alzheimer's disease, and more focal impairments are observed.
- Dementia may also be caused by other degenerative diseases, trauma, infection, neurotoxins, and systemic illnesses (Box 15.1).

Clinical Management

- Careful evaluation of causes is essential.
  - About *15% of cases are potentially reversible.*
  - *Neuropsychologic assessment* can be valuable in early identification and in establishing a base line against which to measure later status.
- *Environmental management* includes providing appropriate assistance and supervision and maintaining a stable and consistent environment.
- *Family support* and *counseling* assist the family with understanding the patient's condition and needs, coping with specific problems such as agitation or wandering, and making decisions about nursing home care.
- Many patients can benefit, at least for a time, from structured recreation, socialization, and exercise aimed at maintaining maximal quality of life.

Chapter 15 Cognitive Disorders

Box 15.1

SELECTED CAUSES OF DEMENTIA

Degenerative Disease
Alzheimer's disease
Pick's disease
Multiple sclerosis
Huntington's disease
Parkinson's disease

Other Neurologic Causes
Trauma
Multi-infarct dementia
Cerebral hypoxia
Neoplasm

Infections
AIDS
Encephalitis
Meningitis
Jakob-Creutzfeldt disease

Substances
Alcohol
Inhalants
Solvents

■ Pharmacotherapy

• *Tacrine* may bring some temporary improvement in cognition and functioning for patients with Alzheimer's disease, but adverse effects limit usefulness.

• Depression, anxiety, agitation, and psychotic symptoms may require medication.

• Especially with Alzheimer's disease patients, caution is needed regarding anticholinergic side effects.

AMNESTIC DISORDER

Amnestic disorder is characterized by impairment in memory without the more generalized cognitive impairment of delirium or dementia.

■ *Clinical features* manifest as impaired ability to learn new information or inability to recall previously learned information, leading to impairment in social or occupational functioning.

■ **Causes**

● *Korsakoff's syndrome,* associated with alcohol abuse, is the most common.

• The syndrome is a residual of *Wernicke's encephalopathy,* which results from *thiamine deficiency* and is characterized by confusion, ataxic gait, and nystagmus.

• Even after thiamine replacement and with abstinence from alcohol, 50% to 60% of patients will have some continuing impairment.

● Trauma, neoplasm, infarction, and hypoxia are other potential causes.

■ *Management* consists of assisting patients with compensatory strategies and environmental cues.

MULTIPLE CHOICE REVIEW QUESTIONS

1. In delirium which of the following is *true*?
   a. Onset is insidious and progressive.
   b. Patients tend to show fluctuating levels of consciousness and awareness.
   c. Attention is intact but memory is poor.
   d. Impairment is commonly more severe in the morning.

2. In a patient with pseudodementia which of the following is *true*?
   a. Symptoms result from medication reactions or metabolic disorders.
   b. The condition is reversible.
   c. There is no insight into or concern about memory failures.
   d. Severe depression creates the impression of dementia.

3. Patients with Alzheimer's disease show which of the following?
   a. Gradual and progressive decline in functioning.
   b. Better functioning in a maximally active and unstructured environment
   c. More impairment in remote recall than in recent recall
   d. Impairment of memory but not in other functions

4. Alzheimer's disease is associated with which of the following?
   a. Thiamine deficiency
   b. Degeneration of cholinergic neurons
   c. Multiple cerebral infarctions
   d. Dopamine hypoactivity

5. In Korsakoff's syndrome which of the following is *correct*?
   a. Onset is insidious and course is progressive.
   b. Impairment is usually transient and fully resolves.
   c. The patient likely has a history of alcohol abuse.
   d. The cause may be trauma, infarction, or hypoxia.

Chapter 16

Disorders Associated with Childhood and Adolescence

DEVELOPMENTAL DISORDERS

■ *Mental retardation* is characterized by subnormal intellectual functioning and adaptive deficits.

● **Characteristics**

• Subnormal intellect is defined as an *IQ of about 70 or less* on an individually administered intelligence test.

• Adaptive deficits refer to functioning below age-appropriate levels in skills relevant to self-care, communication, social relationships, work, health and safety, and use of resources.

• Retardation is characterized as *mild, moderate, severe,* or *profound* (Table 16.1).

• *Prevalence is 1% to 2%* of the population, with males being affected more often than females (about 1.5:1).

● **Causes**

• *Causes* include genetic abnormalities, prenatal infections and toxins, and conditions acquired in infancy and early childhood (Box 16.1).

• The *most common causes* are *Down syndrome, spina bifida, fetal alcohol syndrome,* and *maternal rubella.*

■ **Autism (Pervasive Developmental Disorder)**

● **Characteristics**

• Deficits in social relatedness, impaired communication, and restricted and stereotyped patterns of behavior and interests

• Onset before 3 years of age

• Some degree of *mental retardation* observed in 70% of patients with autism

● **Prevalence and etiology**

• Prevalence is 1.0 to 1.5 per thousand, and the disorder is *3 to 4 times more common in boys.*

• Genetic causes are implicated when one finds a concordance of more than one third in monozygotic twins and no concordance in dizygotic twins.

• Neuroanatomic findings include *abnormalities in the cerebellum* and *ventricular enlargement.* A comorbid *seizure disorder is seen in 25% of cases.*

● **Course and treatment**

• *Course is lifelong,* and morbidity is severe; only *2% to 3%* will complete a normal education or live and work independently.

113

Table 16.1 *Levels of Retardation*

Level	IQ Range	Description
Mild	55 to 70	Sixth-grade academic skills Can develop social, vocational, and communication skills for basic self-support
Moderate	35-40 to 50-55	Second-grade academic skills Can perform unskilled or semiskilled work in sheltered setting More limited social skills Needs some supervision
Severe	20-25 to 35-40	Training limited to elementary habits Marginal communication and social skills Requires controlled environment
Profound	Below 20-25	Minimal capacity for self-care May be unable to communicate Needs nursing care and constant supervision

Box 16.1

CAUSES OF MENTAL RETARDATION

Genetic Abnormalities
Down syndrome
Fragile X syndrome
Phenylketonuria
Tay-Sachs disease

Childhood Disorders
Encephalitis
Meningitis
Lead poisoning
Trauma
Hypoxia from near drowning

Prenatal Conditions
Fetal alcohol syndrome
Maternal rubella
Toxemia
Maternal diabetes
Syphilis
Toxoplasmosis
Malnutrition
Herpes simplex
HIV infection

- *Treatment* involves behavioral interventions to improve communication and social functioning and specialized educational programming.
- Low doses of neuroleptics such as haloperidol may help reduce aggressiveness and self-stimulating or self-injurious behavior.

■ *Learning disabilities* are characterized by the inability to achieve in one or more academic areas at a level consistent with intellect.

- Intellect must fall in the normal range.
- If there are specific sensory deficits, the learning problem must be more severe than would be expected on the basis of the sensory deficits alone.

Chapter 16 Disorders Associated with Childhood and Adolescence

- Learning disorders are *2 to 4 times more common among boys* than among girls.
- Potential consequences include academic failure, poor self-esteem, and acting out associated with frustration in the school setting.
- Treatment should include careful assessment, remedial education, and compensatory skills.

Other Disorders Associated with Childhood

■ Attention Deficit and Hyperactivity Disorder (ADHD)

- **Characteristics (Box 16.2).**
    - Difficulty focusing and sustaining attention
    - Hyperactivity and impulsivity

- **Associated phenomena**
    - Learning disorders, mood disorders, substance abuse, and antisocial personality are more common than usual among family members.
    - *Neurologic soft signs* are more common than normal among children with ADHD, and up to 20% show electroencephalogram abnormalities.

- **Prevalence**
    - *3% to 8% of school-aged children*
    - 3 to 5 times more common in boys

- **Etiology**
    - Genetic factors are implicated by familial aggregation.
    - Prenatal and perinatal factors such as maternal substance abuse or illness, malnutrition, and premature or complicated delivery may be contributory.

Box 16.2

SYMPTOMS OF ATTENTION DEFICIT AND HYPERACTIVITY DISORDER
Difficulty Sustaining and Focusing Attention
Easily distracted by extraneous or irrelevant stimuli
Unable to persist in tasks to completion
Careless mistakes and inattention to detail
Difficulty organizing tasks and activities
Does not seem to listen and fails to follow instructions
Forgetful and prone to losing things
Resists or avoids tasks requiring sustained mental effort
Hyperactivity and Impulsivity
Fidgets and has difficulty sitting still
Excessive activity and restlessness
Interrupts and has difficulty waiting in turn
Talks excessively and has difficulty playing quietly
Blurts out comments or answers to questions

- Course
  - There is typically some reduction in symptoms with puberty, but problems with attention and impulse control persist into adulthood in about half of cases.
  - Up to 25% will meet criteria for *antisocial personality disorder* as adults.
  - The disorder is associated with increased risk for substance abuse, arrest, accidents, and school failure.

- Treatment
  - *Psychostimulant* medications such as methylphenidate are used to improve attention and reduce hyperactivity.
  - Behavioral management to control disruptive behavior and counseling to deal with issues of self-esteem and negative social reactions are needed.

Separation Anxiety

- Characterized by *recurrent* and *excessive anxiety* when separated from home and loved ones.
  - There is a persistent and excessive worry about loss of or possible harm to attachment figures.
  - The child resists or refuses to go to school, sleep alone, or tolerate being away from home.
- *Treatment* is typically behavioral and should include careful assessment of potential reasons for school avoidance or fear of being alone.

Conduct Disorder

- Characteristics
  - Pattern of behavior that violates social rules and the rights of others
  - Common problems include aggression, destruction of property, and deceitfulness or theft
  - Associated with increased frequency of attention deficit and learning disorders

- Prevalence
  - Found in up to 10% of males and 2% of females under 18 years of age
  - More common in families with substance abuse, antisocial personality, and problems with parenting

- Course and treatment
  - It progresses to adult antisocial personality disorder *in 25% to 40% of cases.*
  - Onset after 10 years of age (adolescent onset) is associated with better outcome.
  - Treatment primarily involves behavioral methods and family therapy.

Oppositional Defiant Disorder

- Characterized by negativistic, hostile, and defiant behavior but without the violations of law and of the rights of others observed in conduct disorder.
  - Argumentative and inclined to blame others for mistakes and misbehavior

- May be often angry, spiteful, and vindictive
- Often disliked by peers and socially isolated and likely to perform poorly in school
- Symptoms often most prominent with adults and peers the child knows well and thus often not readily apparent in clinical interview

Gilles de la Tourette's Disorder

- **Characteristics**
  - It is defined by the presence of *multiple vocal and motor tics.*
  - *Motor tics* usually precede vocal tics in the onset of the disorder and include such actions as sniffing, tongue protrusion, or blinking.
  - *Vocal tics* may begin as grunts or barks and in about 30% of cases include words, sometimes profanities (*coprolalia*).
  - Temporary control over tics may be possible but often is followed by a period of more intense tic behavior.

- **Epidemiology and etiology**
  - It occurs in *less than half of 1% of the population* and is *three times more common among boys* than among girls.
  - The concordance rate of *50% in monozygotic twins* is five times higher than in dizygotic twins.
  - Abnormalities in *dopamine* transmission are implicated by the usefulness of antipsychotic medications that block dopamine receptors.
  - It appears to have a genetic relationship with obsessive-compulsive disorder.

- **Course and treatment**
  - Onset is in childhood or early adolescence.
  - Most patients experience some reduction in tics as they grow older, and symptoms remit in the third decade in about 20% of cases.
  - Low-dose neuroleptics such as haloperidol are the main tool for treatment.
  - Psychotherapy may be needed to help patients adjust to the disorder and the associated social responses.

Multiple Choice Review Questions

1. All *except* which of the following statements about mental retardation is *correct*?
    a. Mentally retarded individuals exhibit an IQ of 70 or below.
    b. A diagnosis of mental retardation requires finding problems in social adaptation.
    c. Retardation is more common in males than in females.
    d. Mildly retarded individuals can typically achieve a second-grade level of academic skills.

2. Which of the following is *true* of children with autism?
    a. They typically have average to above-average intellect.
    b. They are more likely to be girls than boys.
    c. They commonly have significant limitations in the ability to communicate and relate to others.
    d. They are usually able to complete a normal education and live independently.

3. In attention deficit and hyperactivity disorder which of the following is *true*?
    a. Family members have higher than normal rates of substance abuse and antisocial behavior.
    b. About 1 in 5 school-age children meet criteria for the disorder.
    c. Symptoms diminish and gradually disappear as the child passes through puberty.
    d. Antisocial behavior is rare.

4. A 12-year-old boy gets into fights frequently and was recently caught shoplifting. He was also involved in several episodes of vandalism at his school. These features suggest a diagnosis of which of the following?
    a. Oppositional defiant disorder
    b. Autism
    c. Conduct disorder
    d. Attention deficit and hyperactivity disorder

5. Which of the following is *true* of patients with Tourette's disorder?
    a. They will be unable to exert even brief control over tics.
    b. They are likely to be treated with antipsychotic medications.
    c. They will exhibit strictly vocal rather than motor tics.
    d. They are unlikely to show any improvement after puberty.

Chapter 17

Other Psychiatric Disorders

PERSONALITY DISORDERS

■ **General Features**
- Personality disorders are defined by the presence of *pervasive and inflexible patterns of behavior* that cause subjective distress or impaired functioning.
- Patients commonly view their patterns of behavior and thinking as normal or acceptable (*ego-syntonic*).
    - They lack *empathy* and cannot appreciate how their behavior affects others.
    - They seek treatment most often for depression or substance-abuse disorders or at the insistence of others (e.g., family, employers).
- *Course* is typically chronic, and some characteristics are evident as early as adolescence.

■ The *DSM-IV* identifies 10 specific personality disorders in three broad clusters (Table 17.1).
- *Cluster A* is characterized by *oddness* and *eccentricity,* including the schizotypal, schizoid, and paranoid disorders.
- *Cluster B* is characterized by *dramatic* and *erratic behavior* and *affect,* including the antisocial, borderline, histrionic, and narcissistic disorders.
- *Cluster C* is characterized by *fearfulness* and *anxiety,* including the avoidant, dependent, and obsessive-compulsive personality disorders.

■ **Etiology and Management**
- Genetic, neurophysiologic, and environmental factors have been implicated to varying extents.
- *Individual or group psychotherapy* may be of assistance. *Medications* may be used to treat associated anxiety or depression.
- Personality disorders are diagnosed on *Axis II* and do not preclude concurrent Axis I disorders, such as major depression or panic disorder.

EATING DISORDERS

■ **Anorexia Nervosa**
- Its *essential feature* is deliberate restriction of food intake and consequent weight loss of 15% or more of body weight.
- **Associated phenomena**
    - Patients exhibit *disturbance of body image* in which they can see themselves as overweight even when they are becoming emaciated.

119

Table 17.1 *Personality Disorders*	
DISORDER	CHARACTERISTICS
Cluster A	
Schizotypal	Odd or peculiar thinking, speech, behavior, perception; social and interpersonal deficits
Schizoid	Pervasive detachment from social relationships and restricted emotional expression
Paranoid	Pervasive suspiciousness and distrust of others, holds grudges, and is easily slighted
Cluster B	
Antisocial	Failure to conform to social norms, lawbreaking, irresponsible and impulsive, deceitful and unconcerned with rights of others
Borderline	Instability in mood, self-image, and relationships; impulsivity; suicidal and self-mutilating behavior; emotional reactivity
Histrionic	Excessive emotionality, attention seeking, dramatic style, shallow relationships, suggestible
Narcissistic	Grandiose view of self, need for admiration, lack of empathy, and sense of entitlement
Cluster C	
Avoidant	Social inhibiton and withdrawal related to timidity and fear of rejection; inferiority feelings
Dependent	Excessive need to have others make decisions and assume responsibility, fearful of being on their own
Obsessive-compulsive	Preoccupation with orderliness, perfection, and control; rigid and emotionally constricted

- Patients do not cease to feel hunger and often become preoccupied with food.
- Many patients exercise excessively and abuse drugs such as diuretics or stimulants (diet pills) to reduce weight, control appetite, and sustain energy in the face of malnutrition.
- *Onset* is typically in adolescence or young adulthood, and over 90% of patients are female.
- The disorder is more common among women of higher socioeconomic levels and among high achievers.

- *Medical complications* are primarily those of malnutrition and starvation, and *amenorrhea* is common (Box 17.1). Mortalities have been estimated at 5% to 15%.

Bulimia

- Its *essential feature* is a pattern of *binge eating* followed by *purging*.
  - Purging is accomplished by induction of vomiting or by use of medications such as *laxatives* and *diuretics*.
  - Bulimia is distinguished from anorexia by the *maintenance of normal body weight*.

Chapter 17 Other Psychiatric Disorders

> **Box 17.1**
>
> ### MEDICAL COMPLICATIONS OF EATING DISORDERS
>
> **Anorexia**
> Amenorrhea
> Cold intolerance
> Lanugo
> Cardiac abnormalities
> Constipation
> Osteoporosis
>
> **Bulimia**
> Esophageal erosion and tears
> Dental erosion and caries
> Electrolyte imbalances
> Bowel dysfunction (laxatives)
> Salivary gland enlargement
> Dehydration

- *Medical complications* of bulimia are the mechanical consequences of purging (e.g., esophageal erosion) and biochemical effects (e.g., hypocalcemia) (Box 17.1).

Clinical Management of Eating Disorders

- The immediate goal is usually to establish adequate nutritional status with the emphasis on *behavioral techniques* to alter abnormal eating behaviors. Individual, group, and family therapy have application in resolving contributory stresses.

- Medications
  - *Antidepressants* can decrease binging and purging and are useful for co-morbid depression.
  - *Cyproheptadine* helps some anorexic patients gain weight.
  - *Antianxiety* and *antiemetic medications* have application in dealing with distress during the acute phase of reforming eating behavior.

Somatoform and Factitious Disorders

Somatoform Disorders

- The *essential feature* is *physical symptoms* beyond what can be explained by medical findings.
  - Physical complaints are not deliberately feigned or manufactured.
  - Somatoform disorders do not preclude existence of genuine organic diagnoses; symptoms may appear as exaggerated.
- DSM-IV categories (Box 17.2)
- Clinical management
  - Patients are encouraged to maintain a stable relationship with a sympathetic primary care physician to ensure continuity of care and obviate *doctor shopping.*
    - Brief but more frequent visits can be scheduled.

Box 17.2

DSM-IV SOMATOFORM DISORDERS

Somatization Disorder
Multiple, often vague somatic complaints persisting for several years and associated with impaired functioning and excessive seeking of medical evaluation and treatment

Conversion Disorder
Deficit or deficits in voluntary sensory or motor functions suggestive of physical or neurologic cause; often lacking in appropriate concern (*la belle indifférence*)

Hypochondriasis
Excessive and unrealistic preoccupation with fear of having a serious illness; normal variations and minor symptoms are misinterpreted

Pain Disorder
Pain complaints for which psychologic factors play a major role in onset, severity, and persistence and which cause significant impairment or distress

Body Dysmorphic Disorder
Preoccupation with an imagined physical defect or exaggeration of a minor defect to an extent that causes significant distress or social impairment

Undifferentiated Somatoform Disorder
One or more unexplained or exaggerated somatic complaints lasting at least 6 months but not meeting criteria for other somatoform disorders

---

- Confrontation is rarely useful, and the emphasis is on reassurance, conservative treatment, and restoration of activity.

- **Factitious Disorder and Malingering**
  - Factitious disorder (Munchausen's syndrome)
    - It involves *deliberate simulation* of physical or psychological symptoms in the absence of obvious incentive.
    - Patients place themselves at significant risk from strategies employed to produce symptoms (e.g., surreptitious use of medications) and from unnecessary evaluative procedures and treatments.
    - *Munchausen by proxy* refers to simulation of symptoms in another. Most commonly a parent induces or simulates illness in a child.
  - *Malingering* involves deliberate and conscious production of false or exaggerated symptoms, *motivated by specific external incentives.*
    - Simulated illness may be physical or psychiatric.
    - Incentives may include financial gain, avoidance of prosecution, avoiding work, and so on.

DISSOCIATIVE DISORDERS

- The *essential feature* is disturbance in normally integrated functions of identity, consciousness, and memory. The disorders are believed to be rare, though there

> *Box 17.3*
>
> ### DSM-IV DISSOCIATIVE DISORDERS
>
> **Dissociative Identity Disorder**
> Two or more distinct identities with incomplete awareness of each other, at least some of which recurrently take control of behavior
>
> **Dissociative Amnesia**
> Inability to recall important personal information, usually of a stressful or traumatic nature
>
> **Dissociative Fugue Disorder**
> Sudden and unexpected travel from home with loss of identity and possible assumption of a partial or completely new identity
>
> **Depersonalization Disorder**
> Persistent or recurring, distressing experiences of detachment or unreality focused on the body or the external environment

is some controversy over the incidence of dissociative identity disorder (formerly known as "multiple personality disorder").

- ## DSM-IV Categories (Box 17.3)
- ## Clinical Management
  - Nonpsychiatric causes for memory disturbances and other symptoms must be ruled out. Dissociative amnesia lacks the *anterograde* aspect of organic memory loss and is usually *retrograde.*
  - *Fugue and amnesia are usually transient* and typically associated closely with psychosocial stressors.
    - Symptom resolution tends to be abrupt and complete.
    - Therapy then focuses on improving the patient's ability to resolve or tolerate stressors.
  - *Identity disorder and depersonalization disorder* run a more chronic course.
    - Treatment of identity disorder is often extended and commonly includes hypnosis.
    - Depersonalization disorder may respond to serotonin reuptake inhibitors (e.g., fluoxetine).
  - "Amytal interviews" (amobarbitol sodium interviews) are sometimes employed for diagnostic or therapeutic purposes.

SEXUAL AND GENDER IDENTITY DISORDERS

- *Sexual dysfunctions* are characterized by disturbances of the normal sexual cycle from arousal to orgasm (Box 17.4).
  - Patients may suffer from lack of desire, aversion to sex, inability to achieve or sustain physical arousal, or inability to achieve orgasm.

> **Box 17.4**
>
> ### SEXUAL DYSFUNCTIONS
>
> **Hypoactive Sexual Desire**
> Lack of desire for sexual activity, absence of fantasies
>
> **Sexual Aversion Disorder**
> Avoidance of and distaste for genital sexual contact
>
> **Female Sexual Arousal Disorder**
> Inability to achieve or sustain lubrication or excitement
>
> **Male Erectile Disorder**
> Inability to achieve or sustain erection
>
> **Orgasmic Disorders**
> Delay or absence of orgasm after normal excitement
>
> **Premature Ejaculation**
> Ejaculation occurs with brief or minimal stimulation
>
> **Dyspareunia**
> Genital pain before, during, or after intercourse
>
> **Vaginismus**
> Involuntary spasm of vaginal musculature

- If medical causes for dysfunction are ruled out, specialized *sex therapy* may be called for.
    - A careful sexual history provides the basis for a program of education and specific exercises (e.g., *sensate-focus exercises*).
    - Behavioral techniques and couples therapy are often applicable.
- *Paraphilias* are characterized by arousal in response to inappropriate stimuli.
    - Patients may be aroused excessively or only by specific objects (fetishes), inappropriate persons (pedophilia), or inappropriate acts (exhibitionism) (Table 17.2).
    - The paraphilias may involve risk of harm to self or others (e.g., sadism-masochism, pedophilia) and risk of arrest and incarceration.
    - Onset is typically in adolescence, and patients are usually male. Paraphilias can be chronic and highly refractory to treatment.
    - Treatment emphasizes behavioral methods for redirecting arousal, group therapy, social-skills training, and cognitive therapy.
- *Gender identity disorder* is characterized by strong and persistent identification with the opposite sex and a sense that one's own sexual type is somehow incorrect.
    - *Cross gender identification* is expressed by the wearing of clothing, adopting of social roles, and endeavoring to appear as a member of the opposite sex.
        - Patients often express rejection and disgust toward genital and secondary characteristics of their sex (e.g., breasts, facial hair).
        - Characteristics are typically observed in childhood or adolescence.

Table 17.2 *Paraphilias*

Paraphilia	Focus of Arousal
Exhibitionism	Exposing the genitals or public performance of sexual acts
Pedophilia	Sexual activity with children
Voyeurism	Observing others undressing or engaging in sexual activity
Frotteurism	Touching or rubbing against a nonconsenting person
Sexual masochism	Being beaten, bound, humiliated, or otherwise made to suffer
Sexual sadism	Humiliation or abuse of another
Fetishism	Specific objects such as shoes, garments, or materials like rubber
Transvestic fetishism	Cross-dressing

- Patients may seek full *sex-reassignment surgery,* or hormonal treatment to enhance the appearance or feeling of what they believe to be the correct sex.

    • Good adjustment after surgery is associated with relatively good adjustment before surgery, as evident in viable support systems and emotional and occupational stability.

    • Better outcome is also found in better-educated patients with lifelong histories of cross-gender identification.

- Psychotherapy can help patients take a more adaptive view of their anatomic sex and take some degree of pleasure in their own bodies.

Multiple Choice Review Questions

1. A young woman has a history of rapid changes in mood, a very impulsive suicide attempt, and unstable relationships. Which personality disorder is suggested?
   a. Borderline
   b. Narcissistic
   c. Histrionic
   d. Schizotypal

2. Your patient is an aloof and unemotional individual who seems to have no interest in having friends or interacting with others. Which personality disorder is suggested?
   a. Paranoid
   b. Avoidant
   c. Antisocial
   d. Schizoid

3. Which of the following is *true* of bulimia, but not of anorexia?
   a. Patients may abuse laxatives and diuretics.
   b. Antidepressant medications may be clinically useful.
   c. Normal body weight is maintained.
   d. Onset is typically in late adolescence or early adulthood.

4. The characteristic presentation of hypochondriasis involves which of the following?
   a. Multiple somatic complaints
   b. Concern about having a particular illness or condition
   c. Marked indifference about the unusual symptoms reported
   d. Deficits in voluntary sensory or motor functions

5. The primary feature of dissociative identity disorder is which of the following?
   a. Feelings of unreality or detachment
   b. Loss of memory of personally relevant information
   c. The presence of multiple distinct personalities
   d. Abruptly leaving home and taking on a new identity

PART 5
Social Context

Chapter 18

Health Care Delivery and Economics

PHYSICIANS AND OTHER HEALTH CARE PERSONNEL

Physicians

- There are about 650,000 physicians in the United States.
    - About 5% are *Doctors of Osteopathy (D.O.)*, and the remainder have M.D. degrees.
    - About 1 in 5 physicians are foreign medical graduates, including both foreign nationals and Americans who have studied abroad.
    - Before World War II, only 20% of physicians were specialists, but at present about *80%* are specialists, which typically means specialized residency, fellowship training, and board certification.
    - Although the number of physicians graduated annually has approximately doubled in the past 50 years and surpluses are developing in some specialties (e.g., surgical specialties and internal medicine), many rural and inner city areas remain underserved.

Other Health Care Providers

- In 1900, the ratio of physicians to other providers was 1:1, but the ratio at present is 1:20.
- Other health care providers include nurses, pharmacists, physician's assistants, psychologists, dentists, and a variety of therapists and technicians (Box 18.1).
- Licensing and certification
    - *Licensing* regulates the practice of a profession and is determined by *state law*.
    - Licensing laws define the standards of training, scope of practice, and constitution of the executive agency that carries out the law (i.e., a *board of examiners*).
    - *Certification* involves establishing that a person has the qualifications for a specific title (e.g., certified chemical dependency counselor).
    - Certification may be carried out by the state or by professional bodies.

HEALTH CARE DELIVERY SYSTEMS

Individual and Group Practices

- About two thirds of physicians are in private practice, either alone or in group practices.

Box 18.1

NONPHYSICIAN HEALTH CARE PROVIDER

Nurses
   Nurse clinicians
   Nurse practitioners
   Nurse midwives
Physician's assistants
Dentists
Pharmacists
Clinical psychologists
Neuropsychologists
Medical social workers
Psychiatric social workers
Dietitians
Podiatrists
Physical therapists
Speech therapists
Occupational therapists
Respiratory therapists
Radiology technicians

- In addition to traditional *fee-for-service* practices, individuals and groups may participate in larger networks *(independent practice associations, or IPAs, and preferred provider organizations, or PPOs)* that contract as a group to provide services to a specific group of patients at negotiated rates.

Outpatient Facilities

- Hospital-based outpatient clinics provide both general primary care and specialized services.
    - Some hospitals operate outpatient primary care clinics to reduce pressures on emergency room resources.
    - Clinics provide specialized outpatient services and follow-up examinations for specific conditions (e.g., pulmonary disease, cardiac care, chronic pain).
- Public agencies such as state or county health departments often run clinics for sexually transmitted diseases, immunization, and prenatal care.
- Free-standing *immediate care* or *urgent care* clinics have become increasingly common, providing an alternative to the emergency room for relatively limited problems.
- *Ambulatory surgery centers* provide services for which no overnight stay is required, and patients are discharged after recovering from anesthesia.
- *Community mental health centers* provide services including psychiatric care to patients with mental illnesses and often for substance abuse, family and child problems, mental retardation, and other behavioral health problems.

Hospitals

- The majority of hospitals are *not-for-profit* entities, owned by government, community organizations, religious organizations, or universities.
    - About half of U.S. hospitals are owned or managed by multihospital systems.

Chapter 18 Health Care Delivery and Economics

- *Teaching hospitals* are those that operate multiple residency programs and are engaged in clinical training of medical students.
- *General hospitals* provide a wide range of services, whereas *specialty hospitals* provide services in one medical specialty or for more specific needs (e.g., psychiatry, rehabilitation).
- In *acute care* hospitals the length of stay is usually less than 30 days.

- *Public psychiatric hospitals* are typically owned and operated by *state governments*. States may operate separate facilities for acute care and for longer term or chronic care, or may operate both kinds of programs within their institutions.

Federal Systems

- **Department of Veterans Affairs (VA)**
  - The VA operates about 170 hospitals and over 200 ambulatory care facilities.
  - Services are provided to honorably discharged veterans of the armed services, with some priority given to veterans with service-connected conditions.

- *Department of Defense* facilities include hospitals and clinics providing services for military personnel, dependents, and retirees.

- *Department of Health and Human Services* operates the *Indian Health Service,* a primarily rural system providing care for Native Americans.

- *Health Maintenance Organizations (HMOs)* are a rapidly expanding component of the health care system.
  - HMOs are *managed care plans* that integrate the delivery and financing of health care.
    - HMOs provide comprehensive care for enrollees for a monthly fee *(capitation)*.
    - Care may be provided by staff physicians at facilities owned by the HMO *(staff model)* or by community physicians through contracts with PPOs and IPAs *(group or network models)*.
    - Capitation creates pressures for cost control, and one result is an emphasis on *preventive care.*
  - Two thirds of HMOs are *for-profit* enterprises.
    - Reimbursement and copayment policies create strong incentives for enrollees to use providers within the plan.
    - Strong utilization and quality control systems are typical.

- *Hospices* provide care for *terminally ill* patients, meaning those with a life expectancy of less than 6 months.
  - The emphasis is on *supportive care,* including individual and family counseling, pain control, and other palliative measures.
  - Care may be provided on an inpatient basis, but there is often an emphasis on outpatient care where possible and for as long as possible.
  - Care is usually provided by a multidisciplinary team including physicians, nurses, social workers, counselors, clergy, and volunteers.

- *Nursing homes* provide residential care for persons with enduring disabling conditions.

- *Skilled nursing facilities* serve patients whose conditions require higher levels of ongoing medical care.
- *Intermediate-care facilities* serve patients who require more than residential placement but not the more extensive medical services of skilled nursing facilities.
- Pressures for deinstitutionalization and reduction in long-term psychiatric hospitalization have led to increased placement of psychiatric patients in nursing homes.
- *Medicare* and *Medicaid* provide a substantial part of nursing home funding; one third of Medicaid funding goes to nursing home care for indigent elderly.

■ *Quality management* and *utilization review* are increasingly important activities at all levels of health care delivery.

- *Quality control programs* such as Total Quality Management and Continuous Quality Improvement typically use a team approach to identify and intervene in problem areas.
  - Emphasis is on *processes* rather than on individual providers, targetable objectives, and systematic tracking of activity and outcome.
- *Utilization review* involves ensuring that services are necessary and appropriate.
  - *Authorization* after medical documentation of need and requirement of a *second opinion* provide control over unnecessary procedures.
  - *Concurrent reviews* of ongoing cases are a common approach to verifying that services are consistent with the patient's condition.

FINANCING HEALTH CARE

■ *Fee for service* is the traditional system of billing patients (or their insurance plans) for services such as examinations, tests, and treatment procedures.

■ *Prospective payment* systems provide for predefined levels of reimbursements for conditions by *third-party payers* (e.g., insurance companies).

- Conditions are classified into *diagnosis-related groups (DRGs)* for the purpose of deciding on the level of reimbursement.
- These systems are intended to encourage cost containment by allowing providers to keep the difference when costs are lower than expected and requiring them to absorb the difference when costs are higher.

■ **Insurance**

- **Employer-based insurance plans**
  - About 70% of the population (under 65 years of age) is insured through employer benefit programs.
  - Large employers may offer a selection of plans, one of which is a managed care plan (HMO).
- *Insurance carriers* are either for-profit agencies or not-for-profit agencies providing coverage for individuals and employees of businesses.
  - *Blue Cross/Blue Shield* is a well-known not-for-profit carrier; traditionally Blue Cross covers hospital care and Blue Shield covers physician services.

- - In addition to about 70 Blue Cross/Blue Shield plans, there are nearly 1000 commercial carriers.
  - Plans often base premiums on a *group rate,* which pools the resources of many individuals, only a small number of whom make major demands on the system.
- Insurance plans require participants to pay some part of health care costs and impose some limits on what will be covered (Box 18.2).
  - Specific limits on mental health and substance-abuse coverage are very common.
- *Medicare/Medicaid* are federal insurance plans, funded through taxes, providing services to the elderly, the disabled, and low-income patients.
  - *Medicare* serves persons eligible for Social Security, including the elderly and the disabled and persons with chronic renal failure.
    - *Part A* covers hospitalization, care needed after hospitalization, home health agency services, dialysis, and hospice care.
    - *Part B* covers physician visits and supplies and service ordered by the patient's physician.
    - *Part B* is *optional,* and a premium is charged. Patients covered by Medicare are responsible for a *deductible* and *copayments.*
    - Hospital costs are financed by *DRGs,* whereas coverage under Part B is on a *fee-for-service* basis.
  - *Medicaid* covers health services for low-income patients and is financed by a combination of state and federal monies.
    - Certain basic services are federally mandated, but the scope of services and requirements for eligibility are determined at a state level.
    - Medicaid patients do not incur deductibles or copayments for covered services.
    - Services covered include hospitalization, physician visits, prescriptions, lab tests, home health care, and nursing home care.

Box 18.2

FEATURES OF INSURANCE PLANS

Premiums are the monthly or annual fees paid by patients or employers, or both, for insurance coverage.
Out-of-pocket expenses refer to that portion of costs for which the patient is responsible.
Deductibles refer to an amount of expenses the patient must incur (out of pocket) before reimbursement is possible.
Copayments refer to specific amounts or portions of the cost of a service for which the patient is responsible.
Annual maximums represent the financial limit on coverage of the patient's costs in a given year.
Lifetime maximums represent the limits on reimbursable expenses over a given patient's lifetime.

■ *Cost shifting* is a strategy for dealing with unreimbursed and underreimbursed care, primarily through higher charges to available payers.

- A portion of what is commonly charged to private insurance offsets the cost of care for which hospitals are not reimbursed.
- Internal cost shifting allows providers to fund necessary but unprofitable services with revenues from profitable services.

Chapter 18 Health Care Delivery and Economics

MULTIPLE CHOICE REVIEW QUESTIONS

1. Which of the following statements about physicians in the United States is *correct*?
   a. Nearly half of American physicians are foreign medical graudates.
   b. There are about 20 nonphysician health care providers for each physician.
   c. About half of all physicians are specialists.
   d. About 1 in 4 is a Doctor of Osteopathy.

2. Capitation refers to which of the following?
   a. Maximum financial liability and insurance company will assume for a patient
   b. Charging patients a specific per-visit fee rather than a percentage
   c. Paying providers a set fee per patient to provide care
   d. Limitations on maximum physician income

3. Which of the following statements is *correct*?
   a. Hospice programs generally provide services to patients who have a life expectancy of no more than 6 months.
   b. Health maintenance organizations are a nonprofit alternative to traditional insurance.
   c. Utilization review tracks whether insurance companies authorize adequate service.
   d. Federal law prohibits expending Medicaid funding on nursing home care.

4. Which of the following is an example of prospective payment systems?
   a. Fee for service
   b. Hospice care
   c. Preferred provider organizations
   d. Reimbursement by diagnosis related groups

5. All *except* which of the following statements about Medicare and Medicaid is *correct*?
   a. Medicaid patients are not charged a deductible.
   b. Medicare provides for the disabled and patients with chronic renal failure.
   c. Medicaid provides care for low-income patients.
   d. Patients on Medicare must purchase separate coverage for hospitalization.

Chapter 19

Legal and Ethical Issues

CONFIDENTIALITY

- *Confidentiality* refers to the physician's professional obligation to respect the patient's privacy and share information about the patient with others only under appropriate circumstances.
  - Information can generally be shared with other treatment personnel and with clinical supervisors without seeking specific permission.
  - Release of information otherwise requires specific permission from the patient.
    - If the patient is a minor or is incompetent, permission must be obtained from the guardian.
    - Permission is generally specific in that future releases, even to the same party, should be separately authorized by the patient.
    - Written permission is best, but release of information with oral permission is acceptable. Circumstances should be documented.

- **Exceptions**
  - *Duty to warn* applies when a clinician has reason to believe that a patient intends to kill or injure someone. The clinician is obligated to warn the intended victim and the authorities. This duty emerged from a California state court decision in 1976 referred to as the *Tarasoff decision.*
  - **Mandatory reporting**
    - All states require physicians to report suspected physical or sexual abuse of children.
    - Another common state mandate includes reporting abuse of the elderly.
  - Breach of confidentiality may be required by court order *(subpoena)* to appear and provide information in court proceedings or in deposition.
  - Release of minimum necessary information is acceptable in emergency situations.
  - Ordinary confidentiality issues may not be relevant in court-ordered evaluations. The patient should be informed of limitations on the confidentiality of information provided.
- *Insurance carriers* require information for utilization review, quality control, and related purposes. Release of such information is not compelled, but the patient may be denied services or reimbursement if it is withheld.

- **Privilege**
  - Privileged communications are those for which disclosure cannot be compelled (e.g., those between attorney and client).

- The extent to which physician-patient privilege exists is a matter of state law.
- Some states allow for privileged communication between psychotherapist and client, but no such privilege is recognized in federal courts.
- Physician-patient privilege does not exist in military courts.
- Where privilege exists, the right *belongs to the patient,* and only the patient can waive it.

● **Exceptions**
- Privilege is automatically waived if the patient places the medical or psychiatric condition at issue in court (e.g., suing for psychological damages).
- Privilege does not allow a clinician to withhold information when a patient takes legal action against the clinician (e.g., a malpractice suit).
- Privilege may be limited when hospitalization is an issue and in cases involving child abuse or child-custody hearings.

CONSENT AND INFORMED CONSENT

■ The physician who performs a procedure on a patient without *consent* may be liable to a *criminal charge* of battery (unauthorized touching).

■ *Informed consent* involves the ethical duty of ensuring that patients make decisions with a proper awareness of what is involved (Box 19.1).

● Failure to obtain informed consent is a violation of this duty and makes the physician liable to a malpractice suit.

● Consent must usually be obtained from a guardian or parent if the patient is a minor or is *legally incompetent.*
- State law may permit treatment of minors without parental consent for some conditions (e.g., sexually transmitted diseases).
- Parental consent is not required for *emancipated minors.*

COMPETENCE

■ *Competence* becomes an issue when there is reason to believe that a patient is unable to make judgments, comprehend circumstances, and make rational decisions.

Box 19.1

INFORMATION FOR INFORMED CONSENT
Diagnosis
Proposed treatment
Likelihood of success and potential risks associated with the proposed treatment
Alternative treatments
Likelihood of success and potential risks associated with the alternative treatments
Likely outcome in the event of no treatment

- Declaration of incompetence is a *legal* decision and not a clinical one. It is determined by the courts and can be reversed only by the courts.
    - Procedures vary in different states but generally include right to legal representation and other protections of due process.
    - Courts appoint a conservator or guardian to make decisions and manage affairs for the incompetent adult, who may not enter into contracts (including marriage), dispose of his or her property, or consent to medical procedures.
- A patient cannot be declared incompetent solely on the basis of having a psychiatric disorder. It must be established that the disorder causes the patient to be unable to make reasonable and informed decisions on the matters in question.

Civil Commitment

Purposes and Procedures

- States have procedures for involuntary hospitalization of persons who represent a danger to themselves or others by virtue of mental illness.
- Procedures vary but generally involve certification by one or more physicians that the person presents such a danger and that hospitalization is necessary.
- Patients have the right to judicial hearings, legal representation, and other due process safeguards, and commitment is for a limited period of time.

Treatment Issues

- States vary in the extent to which commitment permits involuntary treatment. In some states it must separately be determined that the patient is not competent to refuse treatment.
- The *right to treatment* is the principle that a patient hospitalized involuntarily is entitled to treatment for the condition.
- The principle of *least-restrictive alternative* holds that treatment should involve the minimum necessary restriction on the patient's liberties.

Advance Directives

- Advance directives allow patients to specify, while competent, how their care should be conducted if they become incompetent.
    - *Living wills* specify the patient's wishes about life-sustaining interventions in the event of terminal illness and incapacity. Living wills are accepted in 47 states and the District of Columbia, but only about *15%* of the population have completed one.
    - *Durable power of attorney* allows patients to designate who is to make decisions for them in the event of incapacity.
        - 48 states and the District of Columbia have laws covering designation of surrogates.
        - Usefulness is heavily dependent on the extent to which patient and proxy have discussed end-of-life issues
- *The Patient Self-Determination Act* of 1991 mandated that facilities receiving Medicare and Medicaid funds must provide patients with information about advance directives

Chapter 19   Legal and Ethical Issues   139

MALPRACTICE

Basic Principles

- A successful malpractice action establishes that a physician was *derelict* in a *duty* to provide care, which *directly* caused *damages* (the *Four D's*).

    - Malpractice falls under *tort* law, which involves acts that are not criminal but allegedly wrong another person.

    - If not settled out of court, the action is resolved in a *civil* lawsuit tried in court, with the allegedly wronged patient as the *plaintiff* and the physician as *defendant*.

- The level of proof required of the plaintiff is *preponderance of the evidence* (more likely than not) and *not* the *beyond-a-reasonable-doubt* standard of criminal trials.

- Damages awarded are usually *compensatory*, covering medical costs, lost income, pain and suffering, and other losses. *Punitive* damages may be awarded as well.

- Physicians may be sued over the actions of personnel under their direct supervision, under the doctrine of *respondeat superior*.

Important Issues

- Costs of malpractice insurance are reflected in about 15% of physician fees.

- Malpractice issues may contribute to the practice of *defensive medicine* and may discourage some physicians from high-risk procedures.

- Apart from malpractice, physicians may be sued over other matters, such as financial or contractual disputes, sexual misconduct, or breach of confidentiality.

- State laws, so-called *Good Samaritan statutes,* allow physicians to render care in emergency situations (outside the usual practice settings), with more limited liability.

CRIMINAL RESPONSIBILITY

- *The Insanity Defense* reflects the principle expressed in state laws that an individual may not be culpable for actions by virtue of mental impairment.

    - Specific standards vary but usually involve the idea that *by virtue of mental disease or defect* the person was unable to appreciate the wrongfulness of the act or unable to conform his or her conduct to the requirements of the law.

    - The presence of mental illness is not in itself sufficient to establish that an individual was *legally insane* at the time of an offense.

    - Although controversial, the insanity defense is actually employed in only a very small percentage of criminal cases and is not routinely successful.

Competence

- Independent of the insanity issue, a person must be *competent to stand trial,* meaning able to understand the charges and participate in his or her defense.

- If found incompetent, the person is typically committed for treatment and may be tried later if he or she is determined to be competent.

Multiple Choice Review Questions

1. Which of the following statements about privilege and confidentiality is *correct*?
   a. During legal proceedings, either the patient or the physician can voluntarily waive privilege and allow psychiatric treatment information to be released.
   b. A psychotherapist cannot be compelled to testify about a patient in federal court.
   c. A psychiatric patient who makes his or her mental status an issue in litigation cannot insist on keeping psychiatric records confidential.
   d. A psychotherapist is obliged to protect a patient's privacy even if he or she threatens harm to another person.

2. All *except* which of the following statements about informed consent is *correct*?
   a. A physician who fails to obtain informed consent can be charged with battery.
   b. Informed consent cannot be obtained from someone who is legally incompetent.
   c. Proceeding without informed consent opens the physician to a malpractice suit.
   d. An emancipated minor can give informed consent for health care.

3. When a patient is declared incompetent which of the following is *true*?
   a. The patient is automatically hospitalized until a hearing can be held.
   b. The patient must complete a living will.
   c. The patient is considered incompetent until a physician certifies otherwise.
   d. The patient cannot enter into contracts, including marriage, on his or her own.

4. According to the right to treatment which of the following is *true*?
   a. A patient committed involuntarily cannot refuse treatment, and the state must provide it even against his or her will.
   b. When a patient is committed against his or her will, treatment must be made available for the condition that led to commitment.
   c. The state assumes responsibility for health care for any patient declared incompetent.
   d. States are required by federal law to accept the validity of living wills, and physicians are required to honor their provisions.

5. In a malpractice action which of the following is *true*?
   a. Damages are punitive rather than compensatory.
   b. The standard of proof is less rigorous than in criminal proceedings.
   c. The physician's failure or mistake need not directly cause damage to the patient.
   d. The plaintiff's case must be proved beyond a reasonable doubt.

Chapter 20

Medicine and the Social Sciences

Social and cultural factors influence whether patients perceive themselves to be healthy and involve their actual health status, exposure to risk factors, access to health care, utilization of services, perceptions of health care providers, perceptions that providers have of them, and attitudes and expectations regarding treatment.

Each patient functions in a social environment. Patients are not abstractions, differentiated only by their medical diagnoses, or homogeneous in regard to variables other than their diagnoses. The social sciences (e.g., sociology and anthropology) offer tools and perspectives for enhancing our understanding of that environment and its potential influences.

SOCIOECONOMICS OF HEALTH CARE

Health Care Economics

- Health care costs have been rising as a percentage of the gross national product, reflecting the progress of medical technology, the aging of the population, and the expansion of Medicare and Medicaid.

- The pressures of the aging population are twofold. The elderly account for a disproportionate share of health care costs and are an increasing percentage of the population.

- Estimates vary, but at least 30 million Americans are believed to be without health insurance.

- Although there is a general consensus that everyone should have access to health care, economic realities spur intense debates over how to fund services, what services to fund, and who should be eligible under what circumstances.

■ *Socioeconomic status (SES)* is determined by the interrelated variables of income, education, and occupation.

- **Poverty and health**

    • Persons of lower SES are less likely to have a family physician and more likely to receive care through clinics and hospital emergency rooms.

    • Poor persons are less likely to receive or seek preventive care and less likely to identify symptoms as indicators of the need to seek care.

    • Poverty is associated with reduced life expectancy and higher risks for cancer, heart disease, diabetes, obesity, hypertension, and other chronic and communicable illnesses.

- Risks are attributable partly to the lack of resources and awareness regarding healthful life-styles and features of the physical environment, such as overcrowding and poor hygiene.
- **SES and psychiatric disorders**
  - Epidemiologic studies have indicated that lower SES is associated with *increased risk for psychiatric disorders.*
  - Diagnoses of severe psychiatric disorders, such as schizophrenia, are more common in lower SES populations.
- **Explanations for the health correlates of poverty**
  - The downward-drift hypothesis is based on the suggestion that impaired persons "drift" into lower social classes because they are less capable of effective social adaptation.
  - The *culture of poverty* is a perspective that the environment of poverty encourages and supports unhealthful attitudes and behaviors.

ILLNESS AS A SOCIAL PHENOMENON

- *Illness behavior* refers to the emotional and behavioral responses to illness. What it means to be sick and how one is supposed to behave are partly determined by influences of family and culture.
  - *Emotional reactions* reflect the meanings persons attach to illness.
    - A person's sense of competence, independence, masculinity, or femininity may be affected.
    - Feelings of guilt, rational or not, may occur when patients attribute their conditions to their own behaviors (e.g., smoking, failing to have preventive checkups).
  - *Behavioral responses* involve the way people deal with and interpret symptoms and the ways in which they actualize the patient role.
    - People differ in the extent to which they view symptoms as indicators of illness and causes for seeking care.
    - Patients differ in seeing physician instructions as dictates to be followed or guidelines to be modified according to their own judgment.
    - Patients differ on dimensions of expressiveness, intensity, and stoicism and restraint in discussing symptoms and distress.
- The *sick role* refers to the position of a person identified as ill with respect to social expectations and allowances.
  - *Elements of the sick role* were described in their classical formulation by *Talcott Parsons*.
    - The patient is not responsible for being sick and is entitled to care.
    - The patient is excused from usual social obligations.
    - The patient should want to get well and should seek and cooperate with care.
  - This traditional formulation is seen as limited in applicability to chronic illnesses, mental illness, and patients of varying cultural background.
- **Stigmatization of Mental Illness**
  - Psychiatric disorders are associated with stereotypes about dangerousness and unpredictability. Efforts to develop community-based living sites for psychiatric patients often encounter resistance from the communities.

- There is more ambivalence socially about applying concepts of the sick role to psychiatric disorders. There is ambivalence about accepting that psychiatric patients are not responsible for their conditions or are not able to exert control.

- Some disorders are particularly controversial and are topics of continuing debate.

    - Definition of alcoholism and other addictions as "diseases" is a subject of challenge.
    - Debates over the legitimacy of multiple personality disorder and recovery of repressed memory, and the use of terms such as "battered-wife syndrome" in criminal defenses may erode the legitimacy of psychiatric diagnoses in the public view.

CULTURAL CHANGE AND DIVERSITY

Recognition of the influence of cultural factors can make the physician more effective and enhance physician-patient relationships, but facts employed without sensitivity can lead to stereotyping. The physician should strive for awareness of the cultural backgrounds of patients while still relating to them as individuals who respond in unique ways to their cultures.

Diversity is characteristic of American culture, where all but Native Americans trace ancestry to ethnic groups originating elsewhere.

- *Mainstream* American culture, though its specific characteristics are debatable, features certain general emphases.

    - Nuclear families and financial independence
    - An ideal of upward mobility and home ownership
    - Moderation in the expression of emotion and sexuality
    - A technologic or physiologic rather than spiritual or supernatural view of illness and treatment

- **Minority Cultures**

    - *African Americans* are a diverse part of American culture, but there are important group characteristics, many of which reflect social disadvantages of long standing.

        - Shorter life expectancy than white Americans
        - Just over half the median income of whites
        - Sharp increase in single-parent families headed by females over the last two decades
        - Higher risks for certain illnesses (e.g., hypertension and sickle cell anemia)

    - *Hispanic Americans* include a diverse array of "subcultures" reflecting various countries of origin.

        - The largest Hispanic groups are those of Mexican origin (concentrated in the Southwest) and Puerto Rican origin (concentrated in the Northeast).
            - Language can create obstacles to communication.
            - Common cultural features include emphasis on the extended family.

- *Asian Americans* are also a diverse part of American society, reflecting various countries of origin.
    - Chinese and Japanese are the two largest groups.
    - Asian cultures tend to locate the "person" in the body rather than in the head, creating different attitudes toward organ transplantation and concepts such as brain death.
    - Asian cultures commonly place greater emphasis on extended families and respect for elders than is typical of mainstream American culture.
- *Native American* cultures are unique if only in that they are indigenous, sharing to some extent in the history of dispossession and racism.
    - Many Amerindian communities suffer from high rates of unemployment, alcoholism (and fetal alcohol syndrome), infant mortality, and suicide.
    - Reservation life can be socially isolating and enhance prominence of native concepts of illness and care in the practices of individuals.
- **Common issues**
    - Stresses of immigration
    - Language barriers
    - Culturally specific ideas about health, illness, and treatment
    - Emphasis on extended family relationships
- *Cultural change* alters the context in which health care occurs and is stressful to some extent by definition in presenting demands for adaptation.
    - Urbanization of American society
    - Computerization and other technologic change
    - Changes in family life associated with increased divorce rates and mobility
    - Economic change and related changes in employment patterns

HEALTH BEHAVIOR

Utilization

- Whether or not a patient seeks care is not only a matter of the presence or absence of symptoms.
    - Patients must at the least be aware of the need for a service and must have access to it.
    - Characteristics of the patients, providers, and community all determine awareness and access.
- *Equity theory* identifies factors of *predisposition, enabling,* and *need.*
    - *Predispositions* include demographic characteristics such as education and personal characteristics such as health beliefs, which influence use of health services.
    - *Enabling* factors include existence of necessary personal and community resources.
    - *Need* factors include individual and social or institutional perceptions about what is required.
- The *health-belief model* identifies four areas of perception as determinants of utilization.

Chapter 20 Medicine and the Social Sciences

- Patients must perceive themselves to be at risk.
- Patients must view the risk as a serious one.
- Patients must view the required behavior (e.g., checkup or treatment) as potentially beneficial.
- Patients must view the service as available.

Education and Health Practices

- Better education is no guarantee of healthful practices but is associated with greater life expectancy and other benefits.
- Some of the benefits of education are a function of related factors such as higher standard of living and greater resources.

"Alternative" Medicine and Related Issues

- The term "alternative medicine" incorporates an enormously heterogeneous array of beliefs and practices, a diversity so great as to preclude making summary judgments.
  - Many persons explore the practices and beliefs of other cultures seeking alternatives to what they believe to be an overly technologic and impersonal approach in conventional medicine.
  - Some alternative approaches may have psychologic benefits, and some may have physical benefits, but most are underinvestigated at best.
  - Risks can include failure to seek beneficial treatment in a timely fashion and adverse effects from some treatments.

Faith Healing

- Studies have shown that a substantial part of the population believes to some extent in the efficacy of religious faith, prayer, or religious healers.
- Some communities have mental health professionals who identify themselves as practicing *Christian counseling* or *Christian psychiatry.*
- Especially given the fact that strong religious faith and practice is associated with better physical and mental health, patient views on this are entitled to a respectful physician response.

- *Folk practices* may be very much a part of some cultures, from Puerto Rican communities in New York City to rural Appalachia. Inquiries about patient use of medication should of course include over-the-counter medications and in some circumstances should also include inquiry into use of "home remedies" or herbal preparations.

- *Medical quackery* can be defined as the promotion or use of diagnostic and treatment procedures that are patently ineffectual or dangerous, or should be recognizable as such by an informed person.
  - The *quack* may be a classical con artist motivated solely by profit or may be sincere but misguided.
  - Regardless of motivation, quackery is dangerous in that patients may be discouraged from seeking potentially efficacious conventional treatment or it may be dangerous in itself.
  - Physicians are wise to cultivate relationships with patients in which they will feel comfortable discussing treatments they have heard of or are interested in.

- Even a brief educational conversation with the patient will be more productive than a simple or abrupt dismissal of the topic.

- Dismissal without explanation can confirm what the patient has heard from the practitioner of quackery, to the effect that orthodox medical practitioners are arrogant, unwilling to hear another viewpoint, and defensive.

- Dismissal can also bruise a patient's feelings, since it may imply that the patient was foolish to give the idea in question any credence at all.

Keeping an Open Mind

- Recognizing that data are often scanty or nonexistent, a cautious approach to making judgments is wise.

- The best approach is to keep an open mind while maintaining rigorous intellectual standards about what constitutes adequate data. In evaluating case histories remember why blind and controlled studies are important. As someone once said, "The plural of anecdote is not data."

Chapter 20 Medicine and the Social Sciences

MULTIPLE CHOICE REVIEW QUESTIONS

1. All *except* which of the following statements about socioeconomic status is *true*?
   a. Persons of lower socioeconomic status are less likely to have a family physician.
   b. Lower socioeconomic status is associated with lower life expectancy.
   c. The downward drift hypothesis asserts that exposure to an environment of poverty has negative effects on health.
   d. Diagnoses of the more severe psychiatric disorders, such as schizophrenia, are more common in persons of lower socioeconomic status.

2. In the traditional conception of the sick role which of the following is *true*?
   a. Being ill is stigmatized as a kind of personal failure.
   b. Patients are expected to be incapacitated and unable to make an effort to get well.
   c. Psychiatric illness is given the same consideration as physical illness.
   d. Patients are excused while ill from social and personal obligations.

3. Among minority cultures, we find that which of the following is *true*?
   a. Features of Asian cultures make people less comfortable with organ transplantation.
   b. African Americans are at higher risk than whites for some illnesses, but life expectancy is not significantly different.
   c. Native American communities share an exceptionally low suicide rate.
   d. There has been a gradual decline in the percentage of African American households headed by a single, female parent.

4. The health belief model includes all *except* which of the following elements?
   a. Patients must perceive themselves to be at risk for a health problem.
   b. Patients must view a proposed treatment as consistent with their ideas of healing.
   c. Patients must view the risk of a health problem as serious.
   d. Patients must view a proposed treatment as potentially effective.

Answers and Explanations to Multiple Choice Review Questions

CHAPTER 1: BIOLOGY OF BEHAVIOR

1. **Answer c:**

    Pedigrees track occurrence of a trait or disorder among members of a family tree.

    a: Concordance compares pairs of individuals for cooccurrence of a trait.
    b: Consanguinity refers to sharing descent from a common ancestor.
    d: Heritability estimates the relative importance of genetic factors in the occurrence of a trait or disorder.

2. **Answer a:**

    The right cerebral hemisphere is specialized for nonverbal aspects of language.

    b: Left hemisphere injuries are reflected in disturbance of verbal expression or understanding.
    c: The basal ganglia are involved in control of motor behavior.
    d: The hippocampus is primarily implicated in memory functioning.

3. **Answer d:**

    Changes in judgment, personality, and social behavior are associated with injury to the frontal lobes.

    a: Lesions of the thalamus are associated with alteration in pain perception and deficit in memory.
    b: Lesions of the parietal lobes are associated with impaired somatosensory perception and neglect.
    c: Lesions of the reticular formation are associated with disturbances in sleep and level of consciousness.

4. **Answer c:**

    Movement disorders are associated with abnormalities in the nigrostriatal dopamine pathway.

    a: Norepinephrine abnormalities are associated with depression and sleep disturbance.
    b: Serotonin is associated with depression, sleep disturbance, and eating disorders.
    d: Gamma-aminobutyric acid is most closely associated with anxiety.

5. **Answer a:**

    Alzheimer's disease is associated with decreased acetylcholine levels.

    b: Antipsychotic medications act by blocking dopamine receptors.
    c: Serotonin reuptake inhibitors are used in treatment of depression.
    d: Monoamine oxidase inhibitors are a class of antidepressants.

CHAPTER 2: DEVELOPMENT OF THE LIFE CYCLE

1. **Answer c:**

    The ego is the rational and executive agency in the psychodynamic model of the structure of personality.

    a: Primary process thinking is nonrational and characterizes unconscious mental life.
    b: The pleasure principle governs the functioning of the id, seeking gratification without concern for rational considerations.
    d: The ego ideal develops in the formation of the superego, a nonrational component of personality.

2. **Answer b:**
   Transference is seen when a person relates to others on the basis of feelings about other relationships, such as those with one's parents.

   a: The oedipal conflict involves attraction to the opposite sex parent during the phallic stage of psychosexual development.
   c: Identification involves incorporating the characteristics of others (e.g., parents) in one's own personality.
   d: Fixation is the result of failure to resolve the challenges of a particular stage of psychosexual development.

3. **Answer d:**
   Classical conditioning pairs a previously involuntary response (nausea) with a new stimulus (the clinic).

   a: Stimulus generalization occurs when a response is seen not only to the original stimulus but also to other, similar stimuli.
   b: Spontaneous recovery is the temporary return of a conditioned response after extinction.
   c: Aversive conditioning suppresses a behavior by associating it with unpleasant consequences.

4. **Answer a:**
   Biofeedback facilitates control over normally involuntary responses by providing information about the response.

   b: Aversive conditioning suppresses behavior by introducing undesirable consequences.
   c: Systematic desensitization involves reduction of anxiety through graduated exposure to the feared stimulus.
   d: Flooding is designed to reduce anxiety by full-strength exposure to the feared stimulus, assisting the patient in use of coping techniques.

5. **Answer d:**
   Persons with an internal locus of control believe that health is largely a matter of their own behavior.

   a: Belief that health is determined by complying with authoritative figures such as physicians is characteristic of an external locus of control.
   b: Patients with poor self-efficacy expectations will lack confidence in their ability to carry out relevant behaviors and achieve desirable outcomes.
   c: Belief that health is a matter of luck is a characteristic of an external locus of control.

Chapter 3: Measurement of Behavior

1. **Answer d:**
   Specificity refers to the proportion of patients who do not have a condition when the test results say they do not.

   a: Face validity refers only to whether the test *appears* to measure what it is supposed to.
   b: Positive predictive value is the probability that positive test results mean the patient has the illness in question.
   c: Sensitivity refers to the proportion of patients who have the condition when the results say they do.

2. **Answer c:**
   The lack of obvious meaning requires the subject to project his or her own motives, needs, and dispositions into interpretations of stimuli.

   a: Validity scales are a common part of objective personality tests.
   b: Mental age is a concept employed primarily in older intelligence tests.
   d: Evaluation of potential skills is the province of aptitude tests.

3. **Answer a:**
   The Halstead-Reitan Battery is a neuropsychologic procedure used to evaluate the effects of an insult to the central nervous system.

   b: Diagnosis of mental retardation is based primarily on results of intelligence tests.
   c: Clinical inventories such as the MMPI or MCMI are used in evaluating personality disorders.

Answers and Explanations

d: Determination of potential in normal subjects is the province of aptitude tests.

4. **Answer b:**
   The median is the value at which half of subjects fall below and half fall above.

   a: The mean is the arithmetic average of the scores.
   c: The mode is simply the most frequently occurring score.
   d: The variance is a measure of how much the observed scores deviate from the mean.

5. **Answer c:**
   Analysis of variance compares the value of the dependent variable (blood pressure) under multiple conditions (different medication dosages).

   a: Chi-square examines differences in the number of persons assigned to different categories.
   b: A t-test would be appropriate if *two* dosages were compared.
   d: Multiple regression would be used for examining the effects of more than one factor (two medications, or medication and something else).

CHAPTER 4: PREGNANCY AND INFANCY

1. **Answer c:**
   Sexuality is individual and may be marked by increased or decreased desire.

   a: Quickening occurs during the second trimester.
   b: Morning sickness is prominent in the first trimester.
   d: Emotional changes, including lability, are not unusual.

2. **Answer d:**
   Isolation and inexperience are risk factors for postpartum depression.

   a: Disturbance only reaches the level of major depression in about 10% of new mothers.
   b: Depression is common but usually fairly mild.
   c: New mothers are vulnerable to mood disturbance for a variety of reasons, and about half will have at least a brief period of dysphoric mood.

3. **Answer a:**
   Crying is differentiated according to the stimulus that causes it, and parents can tell the difference.

   b: Colic is excessive crying between 3 weeks and 3 months of age.
   c: Infant behavior resembles gastrointestinal distress, but no specific problem has been identified.
   d: Normal crying peaks at about 6 weeks of age.

4. **Answer c:**
   Tactile sensory abilities are most fully developed at birth.

   a: Neonates show physical and behavioral responses to painful stimuli.
   b: Newborns in fact respond differentially to human faces.
   d: Newborns respond differently to sweet and noxious odors.

5. **Answer c:**
   Infants become depressed when deprived of mother's voice and facial response.

   a: Infants quickly learn to recognize mother's voice.
   b: Anaclitic depression, marked by failure to thrive, is associated with neglect and inattention.
   d: Lifting the infant to the shoulder tends to produce a calm, alert infant.

CHAPTER 5: CHILDHOOD AND ADOLESCENCE

1. **Answer d:**
   A child's early sentences are simplified, lacking such parts as prepositions.

   a: Mental retardation is usually marked by slow development but normal forms.

b: The preoperational stage takes place later than early speech development.

c: Speech in play is not different from speech in other situations.

2. **Answer b:**
Bowel control typically develops between 18 and 24 months of age.

a: Parallel but independent play is typical of children before age 3 years.

c: Bladder control is not typically established until 30 to 36 months, and usually happens first for the daytime.

d: Children typically place both feet on each step until age 3 or 4 years.

3. **Answer a:**
Egocentricity is characteristic of children in the preoperational stage.

b: The concept of conservation is not acquired until the concrete operations stage.

c: Object permanence is usually acquired by age 2 years, the beginning of the preoperational stage.

d: Sensory experience and physical manipulation are the primary features of the earlier sensorimotor stage.

4. **Answer d:**
Segregation of play groups by sex is typical of school-age children.

a: Formal operational thinking does not emerge until adolescence.

b: Boys are not in general bigger or stronger until adolescence.

c: Systematic problem solving develops during this period of concrete operational thinking.

5. **Answer b:**
Alcohol use is the most common problem and a major contributor to accidental death.

a: The leading cause of death is accidents.

c: Early sexual activity is associated with more limited intellectual ability and academic accomplishment.

d: Adolescence is the stage of formal operational thinking.

CHAPTER 6: EARLY AND MIDDLE ADULTHOOD

1. **Answer d:**
Couples who live together before marriage are more likely to divorce.

a: Abuse is more common among couples who live together before marriage.

b: There is no evidence of better sexual relationships among couples who cohabitate.

c: Poor marital adjustment is more common among couples who cohabitate.

2. **Answer b:**
Generativity is about contributing to society and the future, and in Erikson's theory of development is most commonly expressed in having and raising children.

a: Marriage is more a traditional part of the young adult stage, when achieving intimacy is the developmental challenge.

c: Menopause, while a significant event, is not a feature of achieving generativity.

d: Household tasks are not relevant to generativity.

3. **Answer b:**
About 85% of divorced men will remarry.

a: Infidelity is a common cause of divorce in middle-aged couples.

c: Risk of divorce is highest in the first 5 years.

d: The majority of divorcing couples have children under age 18 years.

4. **Answer d:**
Job loss is associated with increased risk of physical illness, depression, and substance abuse.

a: There are about 400,000 new cases of work-related illness annually.

b: Only 25% of reports include occupation or employment status.

c: It is estimated that 90% of family practitioners encounter work-related illness or injury at least once per week.

Answers and Explanations

5. Answer c:

About 80% of middle-aged adults describe their health as good or excellent.

a: The empty nest syndrome is more common among women who have limited involvement in things outside the home.
b: Chronic illness becomes more common in middle age.
d: The male climacterium is not a distinctive physical event and is more psychologic than menopause.

CHAPTER 7: AGING, DEATH, AND DYING

1. Answer d:

Women are several times more likely than men to be widowed.

a: About 95% of older adults reside in the community, alone, or with family.
b: The poverty rate among the elderly is about the same as that in the general population.
c: On reaching age 65 years, the average man will live to age 80 years and the average woman to age 84 years.

2. Answer a:

Age-associated change in sleep architecture includes less time spent in Stage 3 and Stage 4 sleep.

b: Reduction in activity of choline acetyltransferase is one reason why elderly adults are more sensitive to the effects of anticholinergic medications.
c: The ventricles become larger as persons age and the brain atrophies.
d: Cerebral atrophy includes sulcal widening.

3. Answer c:

The refractory period, during which a male is unable to have another orgasm, increases with age.

a: Reduced vaginal lubrication is not unusual in elderly women.
b: Men commonly take longer to achieve erection in old age.
d: Men commonly take longer to reach orgasm in old age.

4. Answer d:

Alzheimer's disease accounts for about half of all dementias.

a: Only about 15% of dementias are from reversible causes.
b: Pseudodementia is associated with depression.
c: Incidence of dementia increases significantly with age.

5. Answer b:

The first of the stages described by Kübler-Ross is denial.

a: Clinical depression is by no means universal among dying patients.
c: Dying patients who become clinically depressed can be usefully treated with antidepressants.
d: Patients do not necessarily pass through all stages, or experience them in a given order.

CHAPTER 8: DOCTOR-PATIENT RELATIONSHIPS

1. Answer d:

Transference is an unconscious reaction to the physician based on previous experiences and not necessarily on the objective characteristics of the physician.

a: Rapport refers to a sense of mutual understanding.
b: Implicit goals are not necessarily unconscious goals and do not refer to how the physician and patient relate to each other.
c: Deliberate efforts to manipulate the physician's perception are not unconscious, although such efforts could be motivated by transference reactions.

2. Answer b:

Assuring the patient that his experience is understandable and an expectable part of the situation is the essence of validation.

a: This statement illustrates confrontation.
c: This question is an example of recapitulation.
d: These are direct questions for eliciting specific information.

3. Answer a:

Physician estimates of compliance commonly are higher than the rates observed in studies of patient compliance with treatment.

b: No "noncompliant personality" has been identified.
c: Compliance is a function of patient characteristics and features of the illness and the treatment, as well as physician factors.
d: Compliance is less likely when treatment requires an extended period of time.

4. Answer c:

Compliance is more of a problem with chronic than with acute illness.

a: Compliance is less likely in the absence of disturbing symptoms.
b: Achieving compliance with lifestyle changes is particularly difficult.
d: Compliance is adversely affected by more complex treatment regimens.

5. Answer c:

Placebo effects are not limited to symptoms caused by psychologic factors.

a: Patients have shown both objective and subjective improvement in response to placebo prescriptions.
b: Placebo responses are not diagnostic of psychosomatic problems.
d: On average, about one third of patients given a placebo will show a response.

CHAPTER 9: STRESS

1. Answer c:

Primary appraisal involves the level of threat.

a: Evaluation of coping skills is considered an element of secondary appraisal.
b: A stressor can present a level of threat even if it is a subjectively positive event.
d: The ability to use coping resources effectively is an element of secondary appraisal.

2. Answer b:

Sustained demands for resistance to stressors can result in pathology before the actual exhaustion of coping capacity.

a: These diseases are associated with the Resistance phase of the GAS.
c: These diseases are associated with the Resistance phase of the GAS.
d: Life-changing events can have ongoing effects, so there is no reason why diseases of adaptation should be uniquely associated with "hassles."

3. Answer a:

Studies have indicated that it is hostility that puts the individual most at risk.

b: Type A behavior is associated with heart disease.
c: Interpreting stressors as challenges is a feature of hardiness, not Type A behavior.
d: Type A behavior is associated with increased risk of myocardial infarction, particularly among those individuals prone to hostility.

4. Answer b:

Many approaches to relaxation are useful, but none is specifically superior.

a: Reframing involves focusing on the most useful interpretation when a situation can be interpreted in more than one way.
c: Exercise facilitates better resistance to stress.
d: Relaxation reduces arousal and enhances the sense of control over one's own state.

5. Answer d:

Pain behavior can be "rewarded" when it elicits assistance and support and other positive consequences, and is likely to increase when rewarded.

a: Pain behavior includes both nonverbal and verbal expressions of pain.
b: Pain behavior is susceptible to the effects of reinforcement.
c: Nociception is a physiological process, and pain behavior does not relate in a

Answers and Explanations

clear way to the physical phenomena of injury and pain.

Chapter 10: Substance Use

1. **Answer d:**

    With many substances that are abused, users require progressively higher dosages to achieve the desired state of intoxication.

    a, c: Withdrawal syndromes can be marked by physiologic and/or psychological features and are a hallmark of dependency.

    b: Continued use in spite of adverse consequences is a hallmark of substance abuse disorders but is not what is meant by "tolerance" in this context.

2. **Answer d:**

    Young males and light smokers are less often counseled to quit smoking.

    a: Less than 10% of physicians smoke.
    b: Less than 75% of physicians routinely advise smoking patients to quit.
    c: Most physicians do not feel prepared to assist patients effectively in quitting.

3. **Answer c:**

    Opioid overdose is characterized by the triad of coma, pinpoint pupils, and respiratory depression; death is from respiratory failure.

    a: Delirium is marked by impaired consciousness, but not by the symptoms described.
    b: Alcohol withdrawal can include delirium and severe physical symptoms, but not coma and the other symptoms of opioid overdose.
    d: Naltrexone is an opioid antagonist.

4. **Answer c:**

    Prolonged usage is associated with a psychosis similar to paranoid schizophrenia.

    a: Delirium tremens is associated with alcohol withdrawal.

    b: Amphetamines produce increased blood pressure, heart rate, and alertness.
    d: Amphetamine use is marked by increased tolerance.

5. **Answer a:**

    Some users will reexperience the effects of use weeks or months later, when they are not using the hallucinogen.

    b: The amotivational syndrome is seen in chronic marijuana use.
    c: Paranoia can be seen in use of cocaine and amphetamine, and, to a lesser extent, in use of marijuana.
    d: Intoxication delirium is seen with various substances (e.g., cocaine).

Chapter 11: Sleep

1. **Answer c:**

    Deeper stages of sleep diminish in old age and may be virtually absent after the sixth decade.

    a: Infants exhibit REM sleep, spending as much as 50% of sleep time in REM sleep.
    b: Percentage of time spent in REM sleep declines from childhood to adulthood.
    d: Both deep and REM sleep decline in old age.

2. **Answer a:**

    Sleep talking occurs during nonREM sleep and is unrelated to dreaming.

    b, c, d: REM sleep is associated with dreaming, and physiologic features include atonia and erections.

3. **Answer d:**

    Snoring is a common problem for patients with apnea and is associated with difficulty in forcing air through airways.

    a: Apnea occurs in 1% to 5% of the population.
    b: Apnea is associated with obesity and a low-hanging soft palate.
    c: Sustained hypertension occurs in about 50% of cases.

4. **Answer b:**
   Sleep episodes are unusual in the rapid onset of REM sleep.

   a: Patients are usually treated with stimulant medications to keep them awake.
   c: Sleep episodes typically last 15 to 60 minutes.
   d: Patients may exhibit cataplexy with abrupt loss of muscle tone.

5. **Answer c:**
   Children are difficult to wake and very difficult to comfort.

   a: Only 1% of cases persist into adulthood.
   b: Children are not dreaming during an episode.
   d: Children usually fall back into a deep sleep almost immediately.

CHAPTER 12: MOOD DISORDERS

1. **Answer b:**
   Hypomania is marked by symptoms of mania with less severe impairment.

   a: These are symptoms of full-blown mania.
   c: Hypomania is not marked by depressive symptoms.
   d: Increased sleep and negative attitudes are characteristic of depression.

2. **Answer a:**
   Lifetime prevalence of depression is several times higher.

   b: Bipolar disorder has the earlier mean age of onset.
   c: Depression is more common among women.
   d: Prognosis is less favorable for bipolar disorder.

3. **Answer d:**
   Episodes of depression typically take 6 months to resolve without treatment and about 3 months with treatment.

   a: About half of patients who experience a depressive episode will have another.
   b: About 20% will develop chronic problems.
   c: Symptoms are typically more gradual in onset and resolution.

4. **Answer b:**
   Females are more likely than males to attempt suicide.

   a: Suicide risk increases with severity of depression.
   c: Older adults, particularly older males, are most likely to commit suicide.
   d: Although females make more suicide attempts, males more often succeed.

5. **Answer d:**
   It generally takes 2 to 4 weeks for benefits of antidepressants to become evident; benefits are evident to observers sooner than to the patient.

   a: Most antidepressants act by inhibiting reuptake of serotonin or norepinephrine.
   b: Lithium is used in treatment of bipolar disorder.
   c: Anticonvulsants are used as alternatives or adjuncts to lithium in treatment of bipolar disorder.

CHAPTER 13: ANXIETY DISORDERS

1. **Answer a:**
   Patients with panic disorder often avoid being out alone for fear of being unassisted in the event of a panic attack.

   b: Depression is observed in up to 50% of patients with panic disorder.
   c: The first attack is most often spontaneous and unconnected to a precipitating stressor.
   d: The typical panic attack lasts 10 to 30 minutes.

2. **Answer d:**
   Social phobia is characterized by fear of embarrassment or humiliation in public situations.

   a: Agoraphobia consists of fear of being away from home, especially alone.
   b: Generalized anxiety is characterized by constant and unfocused worry.

c: Obsessive-compulsive disorder features obsessional thoughts and compulsive rituals.

3. Answer c:

Generalized anxiety is more common among females.

a: Only about 25% of patients develop panic disorder.
b: Onset is usually before age 30 years.
d: Depression and substance abuse are common complications.

4. Answer d:

Anxiety mounts until discharged by the performance of the compulsive behavior.

a: Obsessional thoughts are experienced as undesirable and alien.
b: Patients usually make some attempt to resist carrying out compulsive acts.
c: Compulsive behaviors (e.g., handwashing) often have a quasilogical link with obsessional thoughts (preoccupation with contamination by germs).

5. Answer b:

Antidepressants of all types have been used with effect for most anxiety disorders, including panic disorder and agoraphobia.

a: Buspirone is similar to antidepressants in having a 2 to 4 week delay in onset of effects.
c: Beta-blockers are used in situational anxiety (e.g., "stage fright").
d: Potential for abuse has been a problem with use of benzodiazepines.

Chapter 14: Schizophrenia

1. Answer c:

Negative symptoms involve the absence of normal behaviors and responses, such as the lack of emotional expression.

a: Hallucinations are most often auditory.
b: Peak age of onset is earlier among males.
d: Symptoms of depression are found in 50% or more of patients with schizophrenia.

2. Answer a:

Remissions do not include complete return to baseline levels of functioning.

b: The typical course is one of multiple acute phases and remissions.
c: The prodromal phase precedes the period of active psychosis.
d: Only about 20% of patients achieve relatively normal and integrated social functioning.

3. Answer b:

Presence of primarily positive symptoms is a good prognostic indicator.

a: Later onset is associated with better prognosis.
c: Spontaneous onset, in the absence of stressors, is associated with poorer prognosis.
d: Prominent negative symptoms are associated with poorer prognosis.

4. Answer b:

Studies of cerebral blood flow and metabolic activity indicate hypoactive frontal lobes.

a: Dopamine hyperactivity is indicated by the effects of antipsychotic medications, which block dopamine receptors.
c: Ventricular enlargement is associated with schizophrenia.
d: Studies of blood flow and metabolic activity have indicated hyperactivity of the basal ganglia.

5. Answer c:

Usually seen in patients in long-term treatment with antipsychotic medications, tardive dyskinesia is a potentially irreversible movement disorder.

a: Catatonia is marked by unusual motor symptoms, but not the abnormal involuntary movement of tardive dyskinesia.
b: Waxy flexibility and echopraxia are features of catatonia.
d: Dependence and withdrawal are not seen with antipsychotic medications.

Chapter 15: Cognitive Disorders

1. **Answer b:**
    Awareness can vary rapidly from stupor to agitation.

    a: Onset is typically rapid (hours to days).
    c: Difficulty in focusing and sustaining attention is characteristic of delirium.
    d: Sundowning, or increased impairment in the later portion of the day, is common in delirium.

2. **Answer d:**
    Pseudodementia is primarily of concern in depressed elderly adults.

    a: Causes of dementia include medications and metabolic disorders.
    b: About 15% of dementias are potentially reversible.
    c: Patients with pseudodementia are more likely to report symptoms and express concern about them.

3. **Answer a:**
    Insidious onset and progressive course are characteristic of Alzheimer's disease.

    b: Patients are easily overwhelmed and benefit from a controlled environment with a modest level of stimulation.
    c: Recent memory is more adversely affected in Alzheimer's disease.
    d: Impairment of memory is accompanied by a variety of other cognitive deficits.

4. **Answer b:**
    Degeneration is found in the cholinergic neurons in the nucleus basalis of Meynert.

    a: Thiamine deficiency is implicated in Wernicke's encephalopathy.
    c: Multiinfarct dementia is a separate condition typically marked by more focal impairment and a stepwise course.
    d: Dopamine hypoactivity is associated with Parkinson's disease.

5. **Answer c:**
    Korsakoff's syndrome is associated with alcohol abuse.

    a: Course is not progressive, and onset occurs after a bout of Wernicke's encephalopathy.
    b: Half or more of patients have persisting impairment.
    d: Trauma, infarction, and hypoxia can cause memory impairment but do not cause Korsakoff's syndrome.

Chapter 16: Disorders Associated with Childhood and Adolescence

1. **Answer d:**
    Mildly retarded individuals typically achieve a sixth-grade level of academic functioning.

    a: Subnormal intellect is defined as an IQ of 70 or below on an individual intelligence test.
    b: Subnormal intellect is only one criterion for mental retardation; there must also be a finding of adaptive deficits.
    c: Retardation is more common among males, at a ratio of 1.5:1.

2. **Answer c:**
    Deficits in social relatedness and communication skills are basic features of autism.

    a: Most children with autism are mentally retarded.
    b: Autism is three to four times more common in boys.
    d: Less than 5% of autistic persons complete a normal education and live on their own.

3. **Answer a:**
    Antisocial personality and substance abuse, as well as mood and learning disorders, are more common than normal among family members of patients.

    b: Prevalence is estimated at 3% to 8% of the school population.
    c: Symptoms tend to diminish during puberty, but some problems persist into adulthood in about half of cases.
    d: About 1 in 4 will meet criteria for antisocial personality as adults.

Answers and Explanations

4. **Answer c:**

    Conduct disorder is characterized by aggression and violation of social rules.

    a: Oppositional defiant disorder differs in that behavior is typically negative and hostile but does not involve rule breaking and violation of the rights of others.
    b: Autism involves a more fundamental inability to have relationships.
    d: Attention deficit and hyperactivity can involve impulsive behavior but not necessarily aggressiveness or destructiveness.

5. **Answer b:**

    The usefulness of antipsychotic medications has suggested that dopaminergic hyperactivity is somehow involved in the disorder.

    a: Brief control of tics may be possible, although tic activity often increases temporarily afterward.
    c: Both vocal and motor tics are seen in Tourette's syndrome.
    d: Most patients improve as they grow older, and about 20% will have remission of symptoms during their twenties.

Chapter 17: Other Psychiatric Disorders

1. **Answer a:**

    Borderline personality disorder is characterized by instability and impulsivity.

    b: The narcissistic personality is consistent in having an exaggerated sense of importance and entitlement.
    c: Histrionic personality disorder involves emotionality and a dramatic style, but not the severe instability and self-destructive tendencies of borderline personality disorder.
    d: Schizotypal personality disorder features odd or unusual thinking and behavior.

2. **Answer d:**

    Schizoid personality involves pervasive detachment and restricted emotionality.

    a: Paranoia involves features of suspiciousness and hypersensitivity rather than simple detachment.
    b: Avoidant behavior is fueled by fear and timidity rather than simple disinterest.
    c: Antisocial personality involves shallow relationships but no avoidance of others.

3. **Answer c:**

    Criteria for anorexia include failure to maintain normal body weight.

    a: Abuse of laxatives and diuretics is found in both bulimia and anorexia.
    b: Antidepressant medications are of use with both disorders.
    d: Onset of both conditions is in adolescence or early adulthood.

4. **Answer b:**

    Patients with hypochondriasis are preoccupied with concern that they have a serious illness, and are not reassured by normal examination results.

    a: Multiple, vague somatic complaints are characteristic of somatization disorder.
    c: The phenomenon of "la belle indifference" is seen in conversion disorder.
    d: Conversion disorder involves deficits in voluntary functions, often suggestive of neurologic abnormality.

5. **Answer c:**

    Dissociative identity disorder was formerly called multiple personality disorder.

    a: Feelings of unreality are a feature of depersonalization disorder.
    b: Loss of memory is a feature of dissociative amnesia.
    d: Leaving home and changing identity are seen in dissociative fugue disorder.

Chapter 18: Health Care Delivery and Economics

1. **Answer b:**

    The ratio of physician to nonphysician providers has increased to 1:20 in this century.

a: Only about 20% of American physicians are foreign medical graduates.
c: About 80% of physicians are specialists.
d: Only about 5% of physicians are Doctors of Osteopathy.

2. **Answer c:**

Capitation reimburses physicians a specific amount per patient rather than paying per service.

a: Coverage may have limits, but this does not define capitation.
b: Patients may be responsible for a specific amount or a percentage, but this does not define capitation.
d: Income limitations are not capitation.

3. **Answer a:**

Hospice programs provide services to the terminally ill.

b: More than half of health maintenance organizations are for-profit enterprises.
c: Utilization review establishes whether care is consistent with the patient's condition and needs.
d: A third of Medicaid expenditures are for nursing home care for indigent elderly.

4. **Answer d:**

Reimbursement by diagnosis related groups involves established levels of reimbursement for care for a condition rather than for the amount of service provided.

a: Fee for service is essentially retrospective, with payment coming after a specific service is provided.
b: Hospice care is a set of services that can be paid for in a variety of ways.
c: Preferred provider organizations involve a selected group of physicians who agree to a defined level of reimbursement, but the payment system is not necessarily prospective.

5. **Answer d:**

Medicare covers hospitalization, whereas coverage for physician visits is optional and involves paying a premium.

a: There are no deductibles or copayments for Medicaid patients.
b: Medicare provides for the disabled and for patients with chronic renal failure in addition to providing coverage for the elderly.
c: Medicaid provides care for low-income patients.

Chapter 19: Legal and Ethical Issues

1. **Answer c:**

The patient who makes his or her health status an issue in litigation waives privilege.

a: Where it exists, privilege can be waived only by the patient.
b: Federal courts do not recognize a psychotherapist-patient privilege.
d: The Tarasoff decision established the clinician's duty to warn in the event of specific threats by the patient to harm someone.

2. **Answer a:**

Touching a patient without consent is battery, but simply obtaining consent may not meet the criteria for informed consent.

b: Informed consent depends on the capacity of the patient to participate in a genuine way.
c: Failure to obtain informed consent represents a dereliction of duty, and the patient's remedy is to seek redress in civil litigation.
d: Minors who have established independence are given adult privileges, including control over their own health care.

3. **Answer d:**

Contracts are not legally binding if the person who enters into them is incompetent.

a: Incompetent patients typically require care but not necessarily the restrictive alternative of the hospital setting.
b: An incompetent patient does not have the legal standing to complete a living will.

Answers and Explanations

    c: A patient is declared incompetent in a court proceeding, and the declaration must be reversed in court.

4. **Answer b:**

   The state cannot deprive a person of his or her liberty for psychiatric reasons and then not provide appropriate psychiatric care.

       a: State laws vary in the extent to which commitment implies a right to impose treatment.

       c: The state is not necessarily responsible for health care; the patient's guardian is responsible, and the patient's funds are used as available.

       d: No federal law requires states to honor living wills, but they are accepted in almost every state.

5. **Answer b:**

   The standard of proof is one of "preponderance of evidence," meaning that the plaintiff's case must be proven as "more likely than not" correct.

       a: Damages are primarily compensatory, although punitive damages may be assessed.

       c: The physician's dereliction must directly cause damages.

       d: The standard of "beyond a reasonable doubt" applies to criminal proceedings and not to the civil proceedings of a malpractice case.

Chapter 20: Medicine and the Social Sciences

1. **Answer c:**

   The downward drift hypothesis asserts that mentally ill persons "drift" into lower socioeconomic status because illness prevents them from being socially successful.

       a: Poor people are less likely to have a family physician and more likely to obtain primary care in clinics or emergency rooms.

       b: Life expectancy is correlated with education, income, and other socioeconomic variables.

       d: The downward drift hypothesis was originally developed to explain the higher incidence of major psychiatric diagnoses in lower socioeconomic groups.

2. **Answer d:**

   The sick role includes a social acknowledgment that the patient cannot be expected to meet responsibilities as would be expected in the absence of illness.

       a: An element of the sick role is that illness is not a person's fault.

       b: In return for the allowances society makes, it is expected that the patient will try to get well and minimize the extent to which allowances and care are required.

       c: Psychiatric illness has been socially stigmatized in a way not generally true of physical illness, and the sick role does not apply either.

3. **Answer a:**

   Asian cultures place more emphasis on the body rather than the head as the "seat of personality" and thus view organ transplantation differently.

       b: Life expectancy is lower among African Americans than among white Americans.

       c: The suicide rate is not unusually low, and often is unusually high, in Native American communities.

       d: The percentage of African American households headed by a single, female parent has increased.

4. **Answer b:**

   The health belief model presents health care utilization as a function of the patient's perceptions about risks and benefits.

       a, c, d: The core elements of the model include the patient's belief that he or she is at risk, that the risk is serious, and that the treatment will be of some use in reducing the risk.

INDEX

Pages with *t* indicate tables.

A

Ability, assessment of, 25-27
Abuse of elders, 59
Accidents, 49
Accommodation, 46
Acetaldehyde, 76
Acetylcholine (ACh), 8-9, 93
Achievement tests, 27
Acquisition, 18, 19
Active patient, 65
Acute care hospitals, 131
Acute stress disorder, 99
A delta (Aδ) fibers, 73
Addiction, 79
Adhesives, 82
Adolescence, 48-49
   cognitive development in, 49
   problems of, 49
   social and emotional development in, 49
Adoption studies, 3
Adrenocorticotropic hormone (ACTH), 71
Adulthood
   early, 51
   family and relationships in, 51-52
   leisure in, 52-53
   menopause in, 53-54
   middle, 51
   parenting in, 52
   religious commitment in, 54
   work in, 52-53
Advance directives, 138
Affect, 91
African Americans, 143
Aggression, 13
Agnosia, 109
Agoraphobia, 98-99
Akathisia in schizophrenia, 106
Alanon, 78
Alateen, 78
Alcohol abuse, 76-78, 77*t*
   effects of, 77
   treatment of, 77-78, 78*t*
Alcohol dehydrogenase, 76
Alcohol hallucinosis, 77
Alcoholics Anonymous (AA), 77
Alcohol-induced persisting dementia, 77

Alcoholism, 4, 76-77, 77*t*
Alcohol-related behavior disorders, 40
Aldehyde dehydrogenase, 76
Alprazolam for anxiety disorders, 100
Alternative medicine, 145-146
Alzheimer's disease, 9, 58, 110
Ambiguous stimuli, 28
Ambulatory surgery centers, 130
Amino acids, 8
Amino acid transmitters, 9
Amnestic disorder, 111, 123
   and alcohol abuse, 77
   causes of, 111
   clinical features of, 111
   management of, 111
Amotivational syndrome, 81
Amphetamines, 80
Anaclitic depression, 43
Anal stage, 15
Analysis of transference, 16
Anesthetics, 79
Angel dust, 82
Anhedonia, 92
Anorexia nervosa, 119-120
   clinical management of, 121
Antianxiety agents
   for anxiety disorders, 100
   for eating disorders, 121
Anticholinergic effects in schizophrenia, 106
Anticonvulsants, 79
   for mood disorders, 95
Antidepressants
   for anxiety disorders, 100
   for eating disorders, 121
   for mood disorders, 94-95
   for sleep disorders, 85
Antiemetic medications for eating disorders, 121
Antihistamines, 9
Antipsychotic medications for schizophrenia, 105-106
Antisocial personality disorder, 4, 116
Anxiety, 97
Anxiety disorders, 97
   biology of, 97
   DSM-IV classification of
     acute stress, 99
     generalized, 99
     obsessive compulsive, 99-100

Anxiety disorders—cont'd
  DSM-IV classification of—cont'd
    panic, 97-98
    phobias, 98-99
    posttraumatic stress, 99
  and fear, 97
  signs and symptoms of, 97
  treatment of
    antidepressants, 100
    behavioral methods, 100
    benzodiazepines, 100
Anxiolytics, 79
Apgar scores, 42
Aphasia, 109
Appraisals, 70
Apraxia, 109
Aptitude tests, 27
Asian Americans, 144
Aspartate, 9
Assessment
  of ability and achievement, 25-27
  neuropsychologic, 29-30
  of personality, 27-29
Assimilation, 46
Atrophy, 56
Attachment, 43
Attention deficit hyperactivity disorder (ADHD)
  associated phenomena of, 4, 115
  characteristics of, 115
  course of, 116
  etiology of, 115
  prevalence of, 115
  treatment of, 116
Attributable risk, 31
Autism (prevasive developmental disorder)
  course and treatment of, 113-114
  learning disabilities in, 114-115
  prevalence and etiology of, 113
Autonomic nervous system, 5
Autonomy, 16
Aversive conditioning, 18
Axons, 5
Axon terminals, 5

B

Barbiturates, 10, 78
Beck, Aaron, 22, 28
Behavior, 11
  biology of, 3-11
    measurement of, 24-34
    psychosocial models of, 13-22
Behavioral biochemistry, 6
  neurotransmission, 7-8
Behavioral disinhibition, 76
Behavioral management of chronic pain, 74
Behavioral rehearsal, 72
Behavioral responses, 142
Behavioral therapy
  for anxiety disorders, 100
  for chronic pain, 74
  for mood disorders, 95
Behavior modification, 20-21
Bell-shaped distribution, 31

Benzodiazepines, 76, 77, 80
  for sleep disorders, 85
Benztropine, 9
Bereavement, 60-61
Beta-adrenergic receptor blockers for anxiety disorders, 100
Bimodal distribution, 31
Biofeedback, 20, 72, 74
Biogenic amines, 8-9
Bipolar disorders, 91, 92
Birth, premature, 41
Birth defects, 44
Blind procedures, 34
Blue Cross/Blue Shield, 132-133
Bowlby, 60
Bradykinesia in schizophrenia, 106
Brain lesions, behavioral correlates of, 7t
Brazelton scores, 42
Bulimia, 120-121
  clinical management of, 121
Buspirone for anxiety disorders, 100

C

Caffeine, 81
CAGE questionnaire, 76-77
California Personality Inventory (CPI), 27
Cancer, lung, 78
Cardiac disease, 77
Cataplexy, 87
Catecholamines, 71
Causality, 47
Caution for anxiety disorders, 100
Central tendency, measures of, 31
Cerebellum, 6
Cerebral cortex, 6
Certification of health care providers, 129
Cesarean section, delivery by, 40-41
C fibers, 73
Childbirth
  by Cesarean section, 40-41
  neonatal deaths, 41
  postpartum reactions, 41
  premature birth, 41
  stillbirths, 41
Childhood. See also Early childhood
  disorders associated with, 113-117, 114t
Chi-square, 33
Choline acetyltransferase, 56
Cholinergic activity, 93
Cholinergic agonists, 93
Christian counseling, 145
Christian psychiatry, 145
Chronic obstructive pulmonary disease (COPD), 78
Chronic pain, 73-74
Civil commitment, 138
Classical conditioning, 17-18
Classical psychoanalysis, 16
Classification, 48
Clitoral erection, 84
Clomipramine for anxiety disorders, 100
Clonidine, 79
Cocaine, 81, 82t

Cognitive behavioral concepts
  maladaptive thinking in, 21-22
  modeling in, 21
  observational learning in, 21
Cognitive disorders
  amnesia
    causes of, 111
    clinical features, 111
    management, 111
  delirium
    clinical features, 108
    clinical management, 109
    risk factors and causes of, 108-109, 109t
  dementia
    clinical features, 109
    clinical management, 110
    pharmacotherapy, 111
    risk factors and causes of, 110
Cognitive reframing, 72
Cognitive therapy
  for anxiety disorders, 22, 100
  for chronic pain, 74
  for depression, 22
  for mood disorders, 95
  for stress management, 72
Cohabitation, 51
Colic, 43
Commissures, 6
Communications, privileged, 136-137
Community mental health centers, 130
Competence, 137-138, 139
Compulsions, 100
Concordance, 3, 5
Concrete operations, 48
Conditional stimulus, 18
Conditioned response, 18
Conduct disorder
  characteristics of, 116
  course and treatment, 116-117
  prevalence, 116
Confidentiality, 136-137
Conflict, 13, 49
Confrontation, 66
Consanguinity, 3
Conscious mind, 14
Consents, 137
Conservation, 48
Contingency management, 20, 74
Continuous Quality Improvement (CQI), 132
Control groups, 34
Controls, 34
Coping skills and resources, 72
Corpus callosum, 6
Correlation, 33
Corticosteroids, 71
Corticotropin-releasing factor (CRF), 71
Cost shifting, 134
Countertransference, 16, 65
Criminal responsibility, 139
Cross gender identification, 124-125
Crossover designs, 34
Cross-sectional designs, 34, 57
Crying, 43

Crystallized intelligence, 57
Cultural change, 144
  and diversity, 143-144
Cyclothymia, 91
Cyproheptadine, 9
  for eating disorders, 121
Cytoarchitectonic, 6

D

Death, and dying, 59-61, 59t
Death instinct, 13
Defense, U.S. Department of, 131
Defense mechanisms, 15, 16, 17t
Delirium
  clinical features of, 108
  clinical management of, 109
  risk factors and causes of, 108-109, 109t
Delusional disorder, 106
Dementia, 57
  clinical features of, 109
  clinical management of, 110
  pharmacotherapy for, 111
  versus pseudodementia, 110t
  risk factors and causes of, 110
Dendrites, 5
Dependent variable, 31
Depersonalization disorder, 123
Depressant, 76
Depression, 92
  major, 91
  masked, 92
  severe, 91
Depressive episodes, 91
  features of, 91t
Deprivation, sleep, 87
Desensitization, 20
  systemic, 20
Detoxification, 77
Developmental disorders
  autism
    characteristics of, 113
    course and treatment, 113-114
    learning disabilities, 114-115
    prevalence and etiology, 113
  mental retardation
    causes of, 113
    characteristics of, 113, 114t
Deviation IQ, 25
Dextroamphetamine, 80
Direct questions, 66
Direct relationship, 33
Discrimination, 18, 19
Diseases of adaptation, 70
Dissociative disorders, 122-123
  clinical management of, 123
  DSM-IV categories, 123
Distribution, 31
  bell-shaped, 31
  Gaussian, 31
  normal, 31
Disturbances, 43
Disulfiram (Antabuse), 78
Divorce, 51-52, 52t

Dizygotic (fraternal) twins, 3
Doctor-patient relationship
   adherence and compliance, 66
   clinical communication, roles and expectations, 65-66
   emotion and behavior in illness, 67-68
   interviewing techniques, 66
Doctor shopping, 121
Dopamine, 8-9
   activity, 93
Dopamine hypothesis and schizophrenia, 105
Double-blind controlled studies, 34, 68
Doubt, 16
Down syndrome as cause of mental retardation, 113
Downward-drift hypothesis, 102
Dreams, 14
Dream work, 16
Drive reduction, 13
Drives, 13, 14
Dubowitz scores, 42
Durable power of attorney, 138
Duty to warn, 136
Dystonia in schizophrenia, 106

E

Early childhood
   language in, 47
   physical development in
     cognitive, 46-47
     growth, 46
     motor skills, 46, 47t
     toilet training, 46
   social and emotional development in, 47-48
Eating disorders
   anorexia nervosa, 119-120
   bulimia, 120-121
   clinical management of, 121
Education and health practices, 145
Ego, 14-15
Egocentric, 47
Ego integrity, 16
Elder population
   abuse of, 59
   changes in later life, 56-57
     in brain, 56-57
     physical, 56
   cognitive and intellectual changes, 57
   death and dying, 59-61, 59t
   health problems, 58
   institutionalization, 58-59
   retirement, 57
   sexuality, 57
Electroconvulsive therapy (ECT) for mood disorders, 95
Electroencephalogram (EEG)
   brain ages on, 57
   in investigating sleep, 84
Electromyography in investigating sleep, 84
Electrooculography in investigating sleep, 84
Emotional reactions, 142
Emotion-focused coping, 72
Empathy, 66
Employer-based insurance plans, 132
Empty-nest syndrome, 54

Endogenous opioids, 10
Endorphins, 10
Enkephalins, 10
Epidemiology, 30-31
Erikson, 51
Erikson's stages, 15-16, 16t
Erogenous zones, 15, 15t
Eros, 13
Ethanol, 76
Ethical issues. See Legal and ethical issues
Etiology and management, 119
Exercise, 53, 72
Expectancies, 21
Exposure therapies, 20
   for anxiety disorders, 100
External locus of control, 21
Extinction, 18, 19
Extrapyramidal effects in schizophrenia, 106

F

Facilitation, 66
Factitious disorder, 122
Faith healing, 145
Family risk studies, 3
Fantasies, 48
Fears, 48, 97
Fee for service, 132
Fetal alcohol effect, 40
Fetal alcohol syndrome, 40, 77
   as cause of mental retardation, 113
Fine motor skills, 46
Fixation, 16
Flashbacks, 82
Flattened affect in schizophrenia, 103
Flooding, 20
Folk practices, 145
Formal operational thinking, 49
Free association, 13, 16
Freud's stages, 15, 15t
Frontal lobe, 6
Fuels, 82
Fugue, 123

G

Gamma-aminobutyric acid (GABA), 9
   and schizophrenia, 105
Gastrointestinal disorders, 77
Gaussian distribution, 31
Gender identity disorder, 124-125
General adaptation syndrome, 70
General features, 119
General hospitals, 131
Generalization, 18, 19
Generalized anxiety disorder, 99
Generativity, 16
Genetics
   adoption studies, 3
   heritability, 3
   influences on behavior, 4-5, 4t
   methods of behavioral, 3
   multifactorial nature of inheritance, 3
   and schizophrenia, 105
   twin studies, 3

Gilles de la Tourette's disorder
   characteristics of, 117
   course and treatment of, 117
   epidemiology and etiology of, 117
Glia, 5
Global transmitters, 9
Glues, 82
Glutamate, 9
Glycine, 9
Good Samaritan statutes, 139
Grand mal seizures, 80
Grief, 60-61
Grief disorder, 61, 61*t*
Gross motor skills, 46

H

Hallucinations
   auditory, 108
   hypnogogic, 86, 87*t*
   hypnopompic, 86, 87*t*
   in schizophrenia, 103
   visual, 108
Hallucinogens, 82
Halstead-Reitan Battery, 29
Harlow, Harry, 43
Health
   and midlife, 54
   and sleep, 87
   and work, 53
Health and Human Services, U.S. Department of, 131
Health behavior, 144-145
Health-belief model, 67, 144
Health care
   financing, 132-134
   seeking, 67
   socioeconomics of, 141-142
   and stress, 73
Health care delivery systems
   hospitals, 130-131
   individual and group practices, 129-130
   outpatient facilities, 130
Health care economics, 141
Health care providers, 129
Health maintenance organizations (HMOs), 131
Heart disease, 78
Heritability, 3
Heroin, 79
Heterocyclic antidepressants for mood disorders, 94
Hispanic Americans, 143
Histamine, 9
Holmes, 70
Hormones, 11
Hospice care, 60, 131
Hospitals, 130-131
Hostility, 71
Hypnogogic hallucinations, 86, 87*t*
Hypnopompic hallucinations, 86, 87*t*
Hypnosis, 72
Hypnotics, 80, 85, 86*t*
Hypomania, 91
Hypothalamus, 71
Hypothesis testing, 31-33
Hypoxia, 82

I

Id, 14
Identical twins, 3
Identification, 16
Identity disorder, 123
Illness as social phenomenon, 142-143
Illness behavior, 68
Imaginary companions, 48
Immediate care or urgent care clinics, 130
Immune functioning, 71
Inappropriate affect in schizophrenia, 103
Incidence, 30
Independent variable, 31
Infant behavior. *See* Neonatal and infant behavior
Infant reflexes, 42*t*
Informed consents, 137
Inhalants, 82
Insight, 16
Insomnia
   causal factors of, 85
   treatment options, 85, 86*t*
Institutionalization, 58-59
Insurance carriers, 132-133, 136
Insurance plans, employer-based, 132
Intellect, 4
Intelligence, 25
Intelligence tests, 25
   Wechsler, 26, 26*t*
Intermediate-care facilities, 132
Internal locus of control, 21
Interpretation, 13, 16
Interrater reliability, 24
Interval schedules, 19
Intimacy, 16
Intoxication, legal, 76
Inverse relationship, 33
In vivo exposure, 20
IQ (intelligence quotient), 25-26

K

Kaufman Brief Intelligence Test (KBIT), 26
Kübler-Ross, Elisabeth, 59-60

L

Labile mood in schizophrenia, 103
Lability, 39
Lamaze classes, 40
Language, 47
Latency stage, 15
Learned helplessness, 94
Learning, 17
   observational, 21
Learning disorders, 4, 48, 114-115
Learning theory
   applications of, 19-21
   classical conditioning in, 17-18
   operant conditioning in, 18-19
Left hemisphere, 6
Legal and ethical issues
   advance directives, 138
   civil commitment, 138
   competence, 137-138
   confidentiality, 136-137

Legal and ethical issues—cont'd
  consent, 137
  criminal responsibility, 139
  informed consent, 137
  malpractice, 139
Legal intoxication, 74
Leisure, 52-53
Lesion-behavior correlations, 6, 7t
Libido, 13
Licensing of health care providers, 129
Life-change units, 70
Limbic system, 6, 71
Lithium carbonate for mood disorders, 95
Liver disease, 77
Living wills, 138
Locomotion skills, 46
Locus of control, 21
Longitudinal studies, 34
LSD (lysergic acid diethylamide), 82
Lung cancer, 78
Luria-Nebraska Battery, 30

M

Magical thinking, 14
Major depression, 41
Maladaptive thinking, 21-22, 72
Malingering, 122
Malpractice, 139
Mandatory reporting, 136
Manic episodes, 91
  features of, 91t
Marijuana, 81
Marriage, 51
Masked depression, 92
Maternal rubella as cause of mental retardation, 113
Maternal smoking, 40
Maternal use
  of alcohol, 40
  of cocaine, 40
  of opiates, 40
MDMA, 80
Mean, 31
Measurement of behavior, 24-34
Median, 31
Medicaid, 132, 133
Medical care. *See* Health care
Medical quackery, 145-146
Medicare, 132, 133
Meditation, 72
Melancholia, 92
Memory deficit, 109
Menopause, 53-54
Mental age, 26
Mental illness, stigmatization of, 142-143
Mental retardation, 48
  causes of, 113
  characteristics of, 113, 114t
Metanalysis, 34
Methadone, 79
Methamphetamine, 81
Methylphenidate, 81
Michigan Alcohol Screening Test (MAST), 28, 76

Middle adulthood
  health in, 54
  issues in, 53-54
Midlife crisis, 54
Million Clinical Multiaxial Inventory (MCMI), 27
Mind
  conscious, 14
  preconscious, 14
Minnesota Multiphasic Personality Inventory (MMPI), 27, 28t, 74
Minority cultures, 143-144
Mode, 31
Modeling, 21
Modulatory transmitters, 11
Monoamine oxidase inhibitors (MAOIs), 95
Monoamine oxidase levels, 56
Monozygotic twins, 3
Mood, 91
Mood disorders, 91, 91t
  clinical features
    course and prognosis of, 92-93
    psychotic symptoms in, 92
    suicide, 93
    variations, 92
  DSM-IV classification of, 91, 92t
  epidemiology, 91-92
  etiology, 93
    drugs, medications, and nonpsychiatric illness, 94
    psychosocial factors, 94
  treatment of
    antidepressant medications, 94-95
    electroconvulsive therapy, 95
    psychotherapy, 95
    stabilizers and antimania agents, 95
Morning sickness, 39
Motivation, unconscious, 13
Motor skills, 46, 47t
Multidimensional inventories, 27-28
Multifactorial nature of inheritance, 3
Multi-infarct dementia, 110
Multiple regression, 33
Munchausen by proxy, 122
Munchausen's syndrome, 122
Muscarinic class of ACh receptors, 9
Muscle atonia, 84
Muscle tone in investigating sleep, 84
Myelin sheath, 5
Myocardial infarction, 71
MZ twins, 4

N

Naloxone, 79
Naltrexone, 79
Narcolepsy, 87
Native Americans, 144
Negatively skewed distribution, 31
Negative predictive value, 25
Negative reinforcement, 19
Neonatal and infant behavior
  assessment of, 42
  attachment in, 43
  colic in, 43
  competence in, 41-42, 42t

Neonatal and infant behavior—cont'd
    crying in, 43
    disturbances in, 43
    neonatal assessment of, 42
    perceptual skills in, 42
    problem infants in, 44
    reciprocity in, 43
    social responses in, 42
    temperament in, 43-44
Neonatal deaths, 41
Neuroanatomy
    behavioral geography, 6
    structural anatomy, 5-6
Neurochemical change, 56
Neurons, 5
Neuropeptides, 8, 10-11
Neuropsychologic assessment, 29-30
Neuropsychologic batteries, 29-30
Neurotoxicity, 82
Neurotransmission, 7-8
Neurotransmitters, 6
Nicotine, 78
Nightmares, 48
Nociception, 73
Nocturnal emission, 84
Nocturnal enuresis, 46
Nonbizarre delusions, 106
Nonparametric tests, 33
Non-REM sleep, 84
Norepinephrine, 8, 97
    and schizophrenia, 105
Normal distribution, 31
No-treatment control group, 34
Null hypothesis, 31-32
Nursing homes, 131-132

O

Object permanence, 46, 47
Observational learning, 21
Obsessive compulsive disorder, 99
Occipital lobe, 6
Odds, 30
Oedipal conflict, 15
One-way analysis of variance tests, 33
Open-ended questions, 66
Operant conditioning, 18-19
Opioids, 78, 80t
Oppositional defiant disorder, 116-117
Optimal relationships, 65
Oral stage, 15
Outpatient facilities, 130
Over-the-counter stimulants, 81
Oxidation, inhibition of, 7

P

Pain, phenomenon of, 73
Pancreatitis, 77
Panic disorder, 97-98
Paraphilias, 124, 125t
Parasomnias, 86-87
Parasympathetic division, 6
Parenting, 52
Parietal lobe, 6

Parkinson's disease, 9
Passive patient, 65
Patient emotions, responding to, 67
Patient Self-Determination Act (1991), 138
Pavlov, Ivan, 17
PCP (phencyclidine), 82
Peabody Picture Vocabulary Test—Revised, 27
Pedigrees, 3
Penile erection, 84
Perceptual skills, 42
Performance tests, 26
Periodic leg movement disorder, 85
Personality
    assessment of, 27-29
    characteristics of, 4
    structure of, 14-15
Personality disorders
    etiology and management of, 119
    general features of, 119, 120t
Phallic stage, 15
Pheochromocytoma, 97
Phobias, 98-99
Physicians, 129
Physiologic arousal, 72
Piaget, Jean, 48
    theory of, 46
Placebo control group, 34
Placebo effects, 68
Play, 47
Pleasure principle, 14
Polypharmacy, 58
Positively skewed distribution, 31
Positive predictive value, 25
Positive reinforcement, 19
Positive transference, 65
Postpartum blues, 41
Postpartum psychosis, 41
Postpartum reactions, 41
Postsynaptic neuron, 7
Posttraumatic stress disorder, 99
Poverty and health, 141-142
Power of attorney, durable, 138
Preconscious mind, 14
Predictive value, 25, 25t
Pregnancy
    developing, 39
    emotional adaptations to, 39
    prenatal care, 40
    sexuality, 39-40
    smoking in, 78
    substance abuse in, 40
    teenage, 40, 49
Premature birth, 41, 44
Premorbid personality, 102
Prenatal care, 40
Preoperational stage, 47
Preschool children, 48
Prevalence, 30
Primary appraisal, 70
Primary-process thinking, 14
Privileged communications, 136-137
Proband, 3
Problem-focused coping, 72

Prodromal changes, 102
Progressive muscle relaxation, 72
Projective tests, 28-29
Proportionality, 60
Proproteins, 10
Prospective payment systems, 132
Prospective studies, 34
Pseudodementia, 58, 92, 109, 110t
    versus dementia, 110t
Psilocybin, 82
Psychoanalytic psychotherapy, 16
Psychoanalytic therapy, 16, 17t
Psychologic hardiness, 71
Psychologic phenomena, 76
Psychologic tests, 74
Psychopathology, 4-5
    assessment of, 27-29
Psychosexual development, 15-16, 15t, 16t
Psychosis, 102
Psychosocial factors in schizophrenia, 105
Psychosocial interventions in schizophrenia, 106
Psychosocial models of behavior
    defense mechanisms, 16, 17t
    levels of consciousness, 13-14
    principles, 13
    psychoanalytic therapy, 16, 17t
    psychosexual development, 15-16, 15t, 16t
    structure of personality, 14-15
Psychostimulant medications, 116
Psychotherapy for mood disorders, 95
Puberty, 48-49
Public psychiatric hospitals, 131
Punishment, 19

Q

Quality control program, 132
Quality management, 132
Quickening, 39

R

Rahe, 70
Randomization, 34
Rapport, 65-66
Reality principle, 14
Reality testing, 14, 102
Recapitulation, 66
Receptive language, 47
Receptor blockade, 7
Reciprocity, 43
Reflection, 66
Reflexes, infant, 42t
Regression, multiple, 33
Regressive behavior, 73
Reinforcement, 19
    schedules of, 19, 20t
Relative risk, 30
Relaxation techniques
    for anxiety disorders, 99
    in stress management, 72, 74
Reliability
    in research design, 33
    of test, 24
Religious commitment, 54

REM sleep, 84
Repertoire, 18
Repression, 14
Research design, 33-34
Respiratory depression, 79
Reticular activating system, 6
Retirement, 57
Retrospective studies, 34
Reuptake, inhibition of, 7
Right hemisphere, 6
Risk, 30
    attributable, 31
    relative, 30
Rorschach Ink Blot Test, 28

S

Schizoaffective disorder, 106
Schizophrenia and related disorders, 102
    course and prognosis of, 104-105, 105t
    delusional, 106
    epidemiology, 102
    etiology
        genetics, 105
        neuroanatomic correlates, 105
        neurotransmitters, 105
    schizoaffective, 106
    schizophreniform, 106
    signs and symptoms, 102-104, 103t
    subtypes of, 104
    terminology, 102
    treatment of, 105-106
Schizophreniform disorder, 106
Secondary appraisal, 70
Secondary-process thinking, 14
Selective serotonin reuptake inhibitors for mood disorders, 94-95
Self-efficacy, 21
Self-talk, 22
Sensitivity, 24, 25t
Sensorimotor stage, 46
Sentence Completion Test, 29
Separation anxiety, 47, 116
Serotonin, 8
    and schizophrenia, 105
Setting, 66
Sex, in workplace, 53
Sex identity, 48
Sexual and gender identity disorders
    dysfunctions, 123-124
    paraphilias, 124, 125t
Sexual dysfunction, 123-124
Sexuality, 13, 39-40, 49, 57
Shame, 16
Shaping, 19
Shipley Institute of Living Scale, 26
Sick role, 67, 142
Silence, 66
Skilled nursing facilities, 132
Skills training, 72
Sleep
    disorders of
        insomnia, 84-85
        parasomnias, 86-87
    and health, 87

Index

Sleep—cont'd
  normal
    age-related changes in sleep architecture, 84
    non-REM, 84
    REM, 84
Sleep apnea, 86
Sleep deprivation, 87
Sleep disorders
  insomnia as, 84-85, 85t
  narcolepsy as, 86
  sleep apnea as, 86
Sleep hygiene, 85
Sleep paralysis, 86, 87t
Sleep terrors, 87
Sleep-wake cycle disturbances, 108
Sleepwalking, 86
Slosson Intelligence Test, 26
Smokeless tobacco, 79
Social and emotional development, 47-48
Social phenomenon, illness as, 142-143
Social phobia, 99
Social Readjustment Rating Scale, 70
Social responses, 42
Socioeconomics of health care, 141-142
Socioeconomic status (SES), 141-142
Solvents, 82
Somatoform disorders, 121-122
Somnambulism, 86-87
Somniloquy, 84
Specificity, 25, 25t
Specific objective measures, 28
Specific phobias, 99
Spina bifida as cause of mental retardation, 113
Spontaneous recovery, 18
Standard deviation, 31
Standardization, 24
Stanford-Binet test, 26, 26t
State-Trait Anxiety Inventory (STAI), 28
Statistical analysis, 33-34
Statistical significance, 32-33
Statistics, 31-33, *32*
Stillbirths, 41
Stimulants, 80-81
Stimulus, 18
Stress
  and chronic pain, 73-74
  defining, 70, 70t
  management of, 72
  and medical care, 73
  modifiers of, 71-72, 72t
  physiology of, 71
Stress inoculation, 72
Stroke, 78
Substance abuse
  alcohol in, 76-77, 77t
  barbiturates in, 79
  basic concepts, 76
  benzodiazepine in, 80
  cocaine in, 81, 82t
  hallucinogens in, 82
  inhalants in, 82
  marijuana in, 81
  opioids in, 79
  stimulants in, 80-81

Substance abuse—cont'd
  tobacco in, 78-79
Substance use, 40, 49
Successive approximation, 19
Suicide, 93
Superego, 15
Support, 66
Supportive care, 131
Surgical anesthesia, 76
Sympathetic division, 5
Synaptic cleft, 7
Systematic desensitization for anxiety disorders, 99
Systemic desensitization, 20

T

Tacrine, 111
Tarasoff decision, 136
Tardive dyskinesia in schizophrenia, 106
Teaching hospitals, 131
Teenage pregnancy, 40, 49
Telegraphic speech, 47
Temperament, 4, 43-44
Temporal lobe, 6
Testing
  predictive value of, 25, 25t
  reliability of, 24
  sensitivity of, 24, 25t
  specificity of, 25, 25t
  standardization of, 24
  validity of, 24
Test of Nonverbal Intelligence (TONI), 27
Test-retest reliability, 24
Tests
  achievement, 27
  aptitude, 27
Tetrahydrocannabinol (THC), 81
Thanatos, 13
Thematic Apperception Test (TAT), 28
Thiamine, 77
  deficiency of, as cause of amnestic disorder, 111
Thought disorder, 102
Thought stopping, 100
Tobacco use, 78-79
Toilet training, 46
Token economies, 21
Topographic theory of mind, 13
Total Quality Management (TQM), 132
Transference, 16, 63
Transient phobias, 48
Tremor in schizophrenia, 106
t-Test, 33
Twin studies, 3
Two-way analysis of variance tests, 33
Type A behavior pattern, 71
Type I error, 32
Type II error, 32

U

Unconditional response, 18
Unconditional stimulus, 18
Unconscious mind, 13
Unconscious motivation, 13
Utilization review, 132

V

Validation in interviewing, 66
Validity
    of research design, 33
    of tests, 24, 25t, 33
Validity scales, 27
Vanillylmandelic acid (VMA), 97
Variability, measures of, 31
Variable, 31
    dependent, 31
    independent, 31
Variance, 31
Variance tests, analysis of, 33
Verbal tests, 26
Veterans Affairs, U.S. Department of, 131
Vulnerable child syndrome, 41

W

Wechsler intelligence tests, 26, 26t
Wernicke-Korsakoff syndrome, 77
Wernicke's encephalopathy, 111
Wide Range Achievement Test-Third edition
    (WRAT-3), 27
Wish fulfillment, 14
Withdrawal delirium, 77, 80
Withdrawal syndrome, 76
Woodcock-Johnson tests of achievement, 27
Work
    and health, 53
    and sex, 53

Mosby's Reviews Series
Copyright © 1996,
Mosby–Year Book, Inc.

How to install this program—Windows users

1. Place the disk in Drive A: (or B:)
2. From Program Manager, select File, then Run, then enter:
   A:SETUP (or B:SETUP if your disk drive is B:)
3. Follow the instructions on screen.

How to run this program—Windows users

Open the MOSBY Program Group and select the ACE program.

How to install this program—Macintosh users

1. Create a new folder on your hard disk called MOSBY, then open it.
   If you already have a MOSBY folder, you may use any other name.
2. Insert the disk into your floppy drive and open it.
3. Drag all items from the disk to the new folder.
4. If more than one disk is included, perform steps 2 and 3 for each disk.

How to run this program—Macintosh users

Open the MOSBY folder and select the ACE program.

For complete instructions on using the program, please read the "How to use this Program" file.

Dedicated to Publishing Excellence

WE WANT TO HEAR FROM YOU!

To help us publish the most useful materials for students, we would appreciate your comments on this book. Please take a few moments to complete the form below, and then tear it out and mail to us. Thank you for your input.

Mosby's reviews: BEHAVIORAL SCIENCE

1. What courses are you using this book for?

__medical school
__osteopathic school
__dental school
__pharmacy school
__physician assistant program
__nursing school
__undergrad
__other_____

__1st year
__2nd year
__3rd year
__4th year
__other_____

2. Was this book useful for your course? Why or why not?

__yes __no _____
_____
_____

3. What features of textbooks are important to you? (*check all that apply*)

__color figures
__summary tables and boxes
__price
__other_____
_____

__text summaries
__self-assessment questions

4. What influenced your decision to buy this text? (*check all that apply*)
__required/recommended by instructor   __bookstore display   __journal advertisement
__recommended by student   ---other _____

5. What other instructional materials did/would you find useful in this course?

__computer-assisted instruction   __slides
__other_____

Are you interested in doing in-depth reviews of our textooks?   ___yes ___no

NAME:_____

ADDRESS:_____

_____

TELEPHONE: _____

THANK YOU!

A Times Mirror Company

BUSINESS REPLY MAIL

FIRST CLASS MAIL   PERMIT No. 135   St. Louis, MO.

POSTAGE WILL BE PAID BY ADDRESSEE

CHRIS REID
MEDICAL EDITORIAL
MOSBY-YEAR BOOK, INC.
11830 WESTLINE INDUSTRIAL DRIVE
ST. LOUIS, MO 63146-9987

NO POSTAGE
NECESSARY
IF MAILED
IN THE
UNITED STATES